Homeowner's Defense

"And How Singularity Will Affect Us in The Future"

By
Arthur St. Thomas

HOMEOWNER'S DEFENSE: The Impact of Singularity on the Future

 This edition incorporates consumer protection, ethical sales psychology, and AI ethics to empower homeowners and citizens in an evolving world.

For permission requests, contact: [Contact Info/Website]

Printed in the USA.

Dedication

Dedicated to those who refuse to be deceived, and who believe Knowledge is the best defense.

About the Author

Arthur is a United States Marine Corps veteran, a seasoned home improvement professional, and a relentless advocate for homeowners. After years of running his own successful contracting business, Arthur made a bold move: he went undercover inside three of the highest-pressure home improvement companies in the country. His mission? To learn exactly how they manipulate customers, inflate prices, and pressure people into signing contracts they don't fully understand. What he discovered was worse than he expected, and it lit a fire. Arthur wrote HOMEOWNERS DEFENSE to expose the tactics used against everyday people and to give them the tools to fight back. He's trained sales teams, built pitch books, and taught objection handling, but now he's using that same knowledge to protect consumers, not exploit them. He believes that justice starts at the kitchen table, with clear information, calm decisions, and the courage to say "no" when something doesn't feel right. To every homeowner, citizen, and consumer who's ever felt pressured, misled, or manipulated. This book is for you. Whether the threat comes from a fast-talking salesperson or a silent algorithm, you deserve truth, clarity, and control. You deserve to understand the tactics used against you and the tools that can protect you. This is your defense manual. Your shield. Your wake-up call. Stay sharp. Stay free.

Foreword

"HOMEOWNER'S DEFENSE" isn't just a book; it's a shield, a training manual, and a wakeup call wrapped in real-world wisdom."

Contents

Contents of this book

Introduction:

Thank you for purchasing my book, "HOMEOWNER'S DEFENSE". You will learn about the industry from my own experiences, and you will also learn many of the ways homeowners are tricked into paying much more for their home improvement projects than necessary. You will learn that nothing is for free, even though companies entice you with fake free offers.

I have also added a lot of information about the singularity and the threat it poses, along with artificial intelligence. The purpose of my book is to inform you about scams pulled on innocent people from all walks of life. Artificial intelligence is going to be a real threat in our future. I once, while watching TV, listened to a news person question a scientist about artificial intelligence. He asked him if AI could ever decide to eliminate humans.

He said computers can only do what we program them to do, so now that can never happen.

Here is the problem. Artificial intelligence has been developed that can teach itself. That means AI can learn on its own beyond what it is programmed for.

Now, many people are worried about this development. So, I've incorporated this information into this book because scammers will also be using this tool to scam people.

When you finish this book, you will know how to avoid being scammed not only by home improvement contractors but from scammers in other businesses.

In 1960, at the age of 17, I joined the USMC and got out in 1965. My only real skill was killing people. Realizing I couldn't make a living doing that, I had to find a way to make a living.

I started working for a small commercial roofing company. In the four years I worked for that company, I learned how to install many types of roofing, sheet metal work, and some carpentry.

When I started my own business, I didn't know anything about sales or advertising. I started getting work through contractors as a subcontractor.

The first year in my own business was good until winter came along. Back then, we used to get severe winter storms, which made it very hard to survive in the roofing business.

That's when I decided I had to learn how to install vinyl siding. (Already knowing how to work with sheet metal made learning this trade much easier.) and get better at carpentry work. Also, I wanted to start selling my own work to make much more money, so I started buying books on salesmanship and advertising methods.

Over the years, I have become very proficient in all aspects of sales, carpentry work, siding, gutters, building additions, and anything else concerning a building or home. What I didn't know was how the con men worked. How did they get the enormous prices for what they sold homeowners? These were the companies that used high-pressure salesmen and subcontractors to do the work on your home. What you will learn in this book is many of the tricks used to con homeowners and consumers of other products as well.

You will learn to spot fake advertisements from these con men and learn why they advertise the way they do. There is also a whole chapter that will put you right at an actual sales event, of how, after going through the training of three of these high-priced companies will show you how it is easy for these people to con you out of your money, causing you to pay a much higher price for your home improvements than you should. Now is your time to fight back and save your money!

CHAPTER ONE
In The Beginning

When I started my own business and started advertising, I noticed ads for roofing and siding that didn't make any sense. These ads were advertising prices for roofing and siding that didn't cover the cost of the material. I knew one of the people advertising this because I had done some of his work as a subcontractor. One day, I ran into him at a restaurant where he and his associates hung out, and I asked him how in the world he could do a job like he was advertising for that cheap. He said, "Kid, all I want to know is who is interested in having that kind of work done. Once I get into their home, I end up getting what I want."

So let me tell you how these five guys operated. Mostly, they would meet at this same restaurant every morning. When they left, they would all go their separate ways.

Their strategy was to go into neighborhoods and knock on doors, trying to get people to listen to them while they tried to sell them a siding job. At that first meeting, he would give them a high price.

If he couldn't get the sale at this very high price, the next day he would come back with one of the other guys in his group. Again, they would knock on these people's doors. He would introduce his friend as his manager, who was there to offer them a better deal. Their reason for this better deal could be that they wanted to use their home as a model home to show off their work.

If these two couldn't get the people to buy at this still high price, they would return with the third guy, who now carried the title of the district manager, to offer them an even better price. And on it went. If they had to use all five guys, they would each have a higher title, and in the end, they usually got the job, at still a high price.

All these old-time con artists operated the same way. Once they sold a siding job, the biggest worry they had was people cancelling the job. The law back then was that once you signed a contract with a contractor, you could cancel it if the contractor hadn't started the job yet. When these con artists sold a job, they would immediately have a couple of boxes of siding with some accessories delivered to the people's house and have one of their subcontractors go there and nail on a few pieces of siding. This was called spiking the job.

Once the job was spiked, you were legally bound to the contract you signed. Your house could sit there for months like that, depending on how many jobs they sold before yours.

Another tool these con men had to work with was being able to provide financing for the people they sold to. There were a lot of loan companies back then who would provide this service to these people and even pay the contractor salesmen points on every loan they secured for them.

Homeowners who financed through these people paid a very high price in interest, which also added to the high cost of the job they were doing on their homes.

The big hook to getting homeowners to buy was that they wanted to use their home as a model home to showcase their work, and would offer them a great deal for letting them use their home for this purpose.

To impress upon you how powerful these con men were, I want to tell you this story before we go into the next chapter.

Some of the con-artist companies didn't just hustle the homeowners; they also cheated the installers out of the money they owed them. This is one incident where my partner and I went and did the job for a contractor. When we finished, we went to his office, and he gave me a check for the amount I gave him. However, the check bounced. When I called his office to ask about it, his brother answered the phone. When I told him the check, they gave me bounced, he gave me the run around, which sounded like I wasn't going to get paid.

I called Ed, who told me about these guys and told him what happened, and that I was going to get my money one way or the other.

Ed told me to be careful how I handle these guys, because they play with guns. On that note, I called my partner, told him what happened, and to come to my house because we were going to get our money.

We had a large axe we used to cut up the barrels of tar into pieces before we put them into the kettle, so I used the grinder to make it really sharp. When my partner asked me why I was bringing the ax, I told him I was told these guys play with guns, I don't have a gun, so I'll use the ax instead.

When we got to their office, the only people there were two secretaries. I asked if their boss was around, and they told me he wasn't but would be soon. I had the ax on my shoulder.

I told them their boss's checks bounced, and if he didn't pay me in cash, they would need a new boss.

When he came in, he rushed into his office, and my partner and I followed him into his office. I laid the ax on his desk, told him his check bounced, that his brother gave me a run around, and if he didn't pay me in cash, I would bury the ax in his head.

He paid, we became good friends, and I did more work for him.

He told me a story about Ed, the man who told me about him. This guy was one of the worst. He would sell people siding jobs, get a big deposit, and sometimes the whole amount, and never go back and do the job. This one incident is where he made a big mistake. This is the story this guy told me. I was sitting at a bar in a hotel in Fall River, Massachusetts, when this guy I'm talking about came in.

After some talk, I asked him about Ed.
He said, "Let me tell you what happened. Some of the boys came up from New York City to see me. While we were driving around Fall River, one of them asked me how Ed was doing. I said, 'Same as always.' As we passed this house, I pointed to it and said, 'You see that house? Ed sold that woman a siding job two months ago, got her to pay the whole amount up front, and never went back to do the job.'
The guy from New York said, 'Stop the car.' Once I stopped, he got out, went into the house, and when he came back, he was mad as hell. I asked why, and he said, 'That's my aunt's house.'
The next morning, around three a.m., four guys broke into Ed's house, dragged him out onto the lawn, pushed him down, and put a gun to his mouth. They asked if he knew that woman at that address, and he nodded yes. The guy from New York told him he had until four o'clock the next day to start and finish the job—or else."
My friend said he didn't know how Ed got the material and crew there so fast, but he did. They worked nonstop until the job was done.
I'm telling you these stories so you'll know just how vicious the home improvement business used to be. Never give a contractor money up front—only pay when he's there with all the materials to finish the job, and even then, only half.
This event happened while I was still working for the first company I started with, before I had my own business. At that time, I had no clue how these con men operated. I didn't even know most of them used subcontractors or how the home improvement system really worked.
Vinyl siding hadn't been introduced yet—aluminum siding was the product of the day. A friend of mine owned a house with peeling paint. I said, "Why don't you put aluminum siding on your house instead of painting it?" He replied, "I'd never put aluminum siding on my house; it cheapens the place because nobody knows what's under it."

Not long after, two sharp operators from New York knocked on his door to sell him aluminum siding. My friend was already deep in debt, and when they backed him into a corner with high-pressure sales tactics, he said, "I like the idea, but I owe too much money every month. There's no way I can buy your siding now."

Did they get the sale? You bet they did. They arranged a loan that paid off all his debts—and the siding job, too. The same man who once swore he'd never put aluminum siding on his house ended up doing it because they made the deal too easy to refuse.

Later in this book, you'll learn why this tactic works so powerfully. Back in those days, these con artists were known as the "blue suede shoe boys." Their time eventually came to an end, but the new generation of con artists became far more sophisticated.

For these high-pressure, high-priced salesmen, the goal was simple: get you to admit the price is too high for you, not for their product. If you agreed to buy right away, fine. But if you said you'd like to have the windows or siding but just didn't have the money right now, that was music to their ears.

Once you gave them that opening, they stopped talking about the high price and started talking about low monthly payments. Once they found out what you could afford, the sale was almost guaranteed.

Back in the day of the blue suede shoe boys, their preferred method was knocking on doors. Their philosophy was simple: by knocking on your door, they eliminated competition. You weren't calling other contractors—they had you all to themselves.

If you invited them in for an estimate, you were in for a high-pressure few hours. Once inside, these guys were nearly impossible to get rid of. That's why they rushed homeowners into signing contracts—because once people had time to think, they'd cancel.

That's why the government eventually created the **three-day cancellation clause** to protect homeowners from high-pressure sales tactics.

Remember this—and I'll remind you again later in the book: some crooked salesmen still won't tell you about those three days. They're required to have your signature on a form explaining your right to cancel, and they must give you a copy of it.

The cancellation clause gives you three days to think about your purchase, and if you decide to change your mind and cancel, you can, only if you sign the contract at your home. The thought there is in your home you can't leave, but in the contractor's office, you can get up and leave. If they start the job on your house before the three days have passed, without your permission, then you will have a legal right not to pay them. Another thing to be careful of.

Sometimes, when we, I mean the companies I was selling for, couldn't get a customer approved for financing from a legitimate loan company, they would take your paperwork to what we used to call a blood bank. These were loan companies that would give any homeowner a loan to improve their homes, no matter how much they already owed other companies, or how bad their credit was.

They didn't really care if you paid them or not because they had a lien on your house, and if you failed to pay them, they wouldn't hesitate to foreclose on you and take your house. Also, the interest they charged was high. If a salesman comes back to you and tells you he couldn't get you approved with the first company but got you approved with another one, be sure to read everything before you sign. It could be the difference between keeping your house and losing it.

Whether you're hiring a contractor, buying a product, or signing up for a service, you're entering a psychological arena. The other side may be trained, rehearsed, and incentivized to win, not by serving you, but by extracting as much as possible from you.

The Battlefield Has Changed

In the past, manipulation came from humans, fast talkers, smooth closers, and pressure-packed pitches. Today, it also comes from machines. Algorithms track your behavior. AI systems predict your decisions. Digital contracts lock you in before you've even read the fine print. This book will teach you how to spot both kinds of manipulation, human and machines, and how to fight back. Your mission starts here. In the chapters ahead, you'll learn how to: Recognize psychological pressure before it hits. Decode the tactics used by contractors, salespeople, and digital platforms. Build your own defense system, one that protects your wallet, your home, and your peace of mind. You don't need to be a lawyer, a tech expert. You just need to be informed, prepared, and willing to act.

Let's begin.

CHAPTER TWO

Sales, Psychology-How Pressure Works

The Sale Begins Before You Know It. Most people think a sales pitch starts when someone opens their mouth. It doesn't. It starts the moment you answer the door, pick up the phone, or click on a website. The pressure is already in motion, subtle, strategic, and psychological.

Sales professionals are trained to influence your emotions before you engage in your logic. They don't need you to think. They need you to feel urgency, trust, fear of missing out, or the desire to be seen as smart, decisive, or responsible.

Pressure Is a Tool -Not Always a Weapon

Let's be clear: pressure isn't always bad. A good coach uses pressure to push you to perform. A firefighter uses pressure to act fast. However, in sales, pressure is often used to bypass your defenses and get you to say "yes" before you fully understand what you're agreeing to.

Here's how it works.

The Core Tactics of Psychological Pressure

Scarcity

"We only have two left." "This deal expires today." Scarcity triggers urgency. It makes you feel like you'll lose something if you don't act now, even if you weren't looking for it in the first place

Authority

"I've been doing this for 20 years."

"This is what all smart homeowners are doing."

People tend to trust perceived experts. The problem is that anyone can sound like an expert with the right tone, uniform, or credentials— real or fake.

Social Proof: "Your neighbors already signed up."

"Everyone's upgrading to this system."

We're wired to follow the crowd. But in sales, "the crowd" is often a script -not a fact.

Reciprocity

"Let me give you a free inspection."

"Here's a discount just for you."

Once someone gives you something, even a small gesture, you feel pressure to give something back -like your time, attention, or trust.

Commitment and Consistency

"You said you wanted to protect your home, right?"

"You already agreed this was important."

Once you say "yes" to something small, you're more likely to say "yes" to something big -even if it's not in your best interest.

Why It Works -And How to Fight Back

These tactics work because they bypass your rational brain. They target your instincts. Your emotions. Your identity.

But once you see them, you can stop them.

Ask yourself: Am I being rushed?

Pause and say: I don't make decisions under pressure.

Flip the script: If this deal disappears today, it wasn't a good deal to begin with.

In the Age of AI, Pressure Gets Smarter

And now imagine these same tactics -but automated. Personalized. Delivered by

an algorithm that knows your browsing history, your fears, your habits. That's not science fiction. That's today's digital sales funnel.

In later chapters, we'll explore how AI uses these same psychological levers -often invisibly - and how to defend yourself in both analog and digital arenas.

In the Age of AI, Pressure Gets Smarter

Your Next Move

But now, you know the game.

And that's the beginning of real defense.

CHAPTER THREE

Scams, Schemes, And Red Flags

If it seems too good, It Probably Is. Scams don't always come wrapped in lies. Sometimes they come dressed as deals. As urgency. As kindness. As "just trying to help." But beneath the surface, they're designed to take your money, your trust, your time.

This chapter is your radar system.

You'll learn how to spot the patterns, decode the setups, and recognize the red flags before they cost you.

The Anatomy of a Scam

Every scam has three core ingredients:

Something to grab your attention, a problem, a fear, or a promise.

"Your roof is leaking."

"This offer won't last."

"I'm here to help."

A Setup

A story that makes the scam feel legitimate.

"We're working in your neighborhood."

"We're certified by the state."

"We helped your neighbor last week."

A Close

A push to act fast, sign now, or pay upfront.

"We just need a deposit."
"Let's lock this in today."
"You don't want to miss this."

Common Schemes to Watch for

The Drive-By Diagnosis:

A contractor knocks on your door and claims they saw damage from the street.

Red Flag: No inspection, no proof, just urgency.

The Phantom Discount

You're offered a "special deal", but only if you sign today.

Red Flag: Real discounts don't expire in hours.

The Credential Mirage. They flash a badge, a logo, or a certificate, but it's outdated, fake, or irrelevant.

Red Flag: Always verify credentials independently.

The Bait-and-Switch

You're quoted one price, then pressured into upgrades or changes mid-project.

Red Flag: Vague contracts, missing details, or verbal promises.

The Vanishing Act

You pay upfront, and they disappear.

Red Flag: No physical address, no references, no paper trail.

Psychological Triggers They Exploit

Fear: "Your home is at risk."
Greed: "This deal is too good."
Trust: "I'm just like you."
Confusion: "Don't worry about the details."

These triggers are designed to short-circuit your judgment. The more emotional you feel, the less rational you become, and that's exactly what they want.

The Digital Twist

Scams aren't just knocking on doors anymore. They're sliding into your inbox, your social media, your smart home.

- Fake contractor websites with AI-generated reviews.
- Deepfake videos of "experts" endorsing products.
- Chatbots posing as customer service reps, guiding you into traps.

The singularity won't just change how we live — it will change how we're targeted. That's why your defense must evolve.

Your Countermeasures

- **Pause:** Scammers hate time. Take it.
- **Verify:** Check credentials, reviews, and references.
- **Document:** Get everything in writing.
- **Ask Questions:** Real professionals welcome scrutiny.
- **Trust Your Gut:** If something feels off, it probably is.

Next Up: Objection Handling — Used Against You

You'll learn how your own words can be turned against you, and how to flip the
script before it's too late. Before we get into that subject, I want to tell you a little about the industry and how I handled finding out exactly how the big hustler companies operated and how they used deceitful tactics to get you to sign on the dotted line.

CHAPTER FOUR

The Beginning of The Replacement Window Industry

For years, the mainstay of the home improvement business was siding, roofing, kitchens, bathrooms, finished basements, and additions. Then, along came one of the biggest money makers in the business, **CUSTOM ENERGY EFFICIENT REPLACEMENT WINDOWS.**

One of the first companies to get into the replacement window business was big on remodeling basements. After getting to know some of their ex-salesmen, I learned their tactic for how they used to sell the basement remodel jobs. When they were pitching to the people, they were trying to sell the remodeling job for their basement, and they would show them many very high-quality paneling samples. Once the homeowner picked the paneling they wanted, the salesman would give them the price of their basement remodel.

The price would be very high. If they couldn't get them to buy at the high price, they would tell them they would be running a big sale on the white paneling the following week, and if they agreed to buy now, they could get them in at the sale price, which was much lower than the first price they got.

Most people, even though they had picked a different panel, went with the cheaper price and the white panel.

What they didn't know was that the owner of the company bought the white paneling by the trailer truckload at a very cheap price, and the idea of giving a very high price at first was to steer the homeowner towards the white paneling. Bait and switch. This former salesman laughed and said we sold tons of white paneling and nobody was the wiser.

As I said earlier, this company was one of the first to enter the window replacement business. The window they sold was an aluminum custom-made window. At this time, I knew nothing about the window business. But one day, the salesmen from my siding supplier came to my job site and said, "Art, we are now making vinyl replacement windows. You must get into this business; it's a big money maker." I said, "I don't know anything about how to install replacement windows." He said, "It's easy, we'll show you how to measure them and how to install them."

It turned out it was very easy to measure and install them. I also learned that the companies selling them at that time were all high-pressure companies with salesmen selling the windows anywhere from five hundred dollars to a thousand dollars apiece. At that time, the average replacement window wholesale was around eighty dollars. Knowing how much these other companies were getting for their windows, I decided to sell mine for three hundred dollars each, making a nice, reasonable profit and not ripping the people off.

Being an amateur window salesman, I would bring in my window sample, open it up, and say, "This is my vinyl window, it is energy efficient, maintenance free and tilts in for easy cleaning. They are three hundred dollars apiece."

I couldn't believe it, using that simple speech, everyone I went to was buying windows from me. Must be they had the high-priced guys there, and my price seemed like a great deal to them. But like in life, all good things must come to an end. On this one call, I went to do my little pitch, and when I told these people the price was three hundred dollars each, he said to me, "Three hundred each, that's awfully high.

I had a guy in here last night with the same window, and he only wanted one hundred and eighty dollars each for them." Well, I didn't get that job. That's when I knew my supplier had too many guys now selling their windows. After thinking about that and realizing how much money could be made selling and installing replacement windows, I decided to shut down my business for one year and go to work for three of the biggest window companies in the state.

I had to learn just how they manage to constantly sell their windows at such a high price and stay in business. So, I managed to get hired by these three companies and go through their training, and go out on their leads to put into practice their techniques to see how they worked.

What you are about to learn in the next chapters will blow your mind.

These people have more ways to have you say yes than you have of saying no. Nobody is safe from these con artists, no matter the reason they are contacting you, no matter what they are trying to sell you; people from all walks of life and all education levels fall for their tactics. You must always be on your guard and very cautious. If your phone rings and it's someone trying to sell you something, and you haven't contacted them first, just hang up. If you don't, you might be getting into something you will regret later.

In the early years, the mode of obtaining your business was newspaper advertising and cold canvassing (knocking on people's doors).

Knocking on people's doors got very difficult because people were getting wiser to these people, and some towns and cities banned these guys from doing it.

That's when the boiler room business got hot. Boiler room operations were companies that would set up many phones in a large room and hire people to cold call homeowners and try to get them to let a salesman come to their homes to tell them about whatever product they were selling.

Back then, the phone companies published a special phone book for boiler room operations. These special phone books listed the people's phone numbers by the street address they lived on, instead of the alphabetical order of their names.

In this way, the cold callers could attack a specific area at a time and not be calling people at random locations.

Now, these calls are made with computers, which can make thousands of calls in a very short time. Robo calls, a nuisance.

Either way, if you agree to let them into your home, you are in for one heck of a trip.

Never agree with anything that wasn't your idea to begin with.

CHAPTER FIVE

Getting Educated-Company One

To find out how the high-pressure companies with salesmen got the very high price for windows, I had to shut down my business for one year and go to work for them to go through their training and learn their sales system.

The first company I went to work for was a Sears authorized company out of New York. I live in Connecticut. The Sears I'm talking about is the Sears chain store that used to be all over the country and old American icon.

Sears would find companies that sold siding and windows that seemed to be solid, with a high expertise in marketing and sales.

This company from New York was given the whole state of Connecticut to use Sears' name and financing to sell their windows. People thought they were dealing with Sears, giving this company a lot of credibility.

For the privilege of using Sears' name and financing, they had to pay Sears fifteen percent of their gross sales off the top. Turned out this company was conning everyone out of their money, the salesmen and Sears included.

I went through their training. After their training, which wasn't very impressive, they sent us out on what were supposed to be good leads.

To get the leads, they used a boiler room operation.

That is, they had a lot of people making cold calls to homeowners trying to set up appointments to have a window salesman come to their house to show them our windows. The leads we're not good leads. Many of the people we went to said they told the person on the phone they weren't interested and couldn't understand why we were there.

I had heard about this scam before. What they were really doing was sending as many people as possible out that they could hire. They didn't really care if we sold anything at all. What they only hoped for was that we got into the house and gave our pitch. The price they wanted us to sell the windows at was so high that we hardly had a chance to sell the windows. This concept was, "If you throw enough crap against the wall, some of it will stick".

The next day, they would find out who got in to pitch people and give them a price. They would then call these people back and ask these questions. Did our salesman show up on time? Did he answer all your questions to your satisfaction? Did you feel comfortable with our salesmen? Did you like our window he showed you? Of course, most people would say yes to all the questions. It's the last question they really wanted a yes to.

Once the person said they liked the window, this person would ask, "If you liked the window so much, why didn't you give our salesmen an order for it?" The answer would always be the same: "It costs too much money."

After hearing that the person from the company would ask, "What if our sales manager could offer you a much lower price? Would you be interested in him coming to your home and talking to you? I'm sure he can save you a lot of money!"

A lot of people agreed, and then the heavy hitter, who was a very experienced high-pressure salesman, would go out and seal the deal. After realizing this company was using us for cannon fodder, I quit and went to work for company number two.

Before we go there, I want to make this point. Unlike the old days, when you signed a contract with a home improvement contractor in your home, you have three days in which to cancel that contract. If you go to his office and sign the contract, then the three-day cancellation clause doesn't apply.

The reason is that when in your home, you can't leave, but when in his office, you can. I'm telling this now because out of all the training I got from the first company, there was only one thing I learned that helped me later, and at the time, I didn't think anything of it, but it later saved me a lot of disappointment. You'll learn about that later.

This company eventually lost its contract with Sears because they weren't paying Sears the proper fifteen percent of total sales.

There was a company from Texas that was so big and sold so many windows using Sears' name that every time Sears dropped a company for the reason, they dropped this company; they would offer this company their territory.

This company got so big, in one year, they sold three hundred and fifty million dollars' worth of windows using Sears' name.

Sears netted fifteen percent of that gross amount of three hundred and fifty million dollars. Not bad for letting someone use your name.

In life, many things aren't what they really appear to be. That's why in the marketing of any business, product deception is used to entice you to become one of their victims.

How do you think politicians get elected? Not by telling us the truth, that's for sure. They're the smartest salesmen in the world. They lie to us every election, and we keep voting for them! When a politician lies and gets caught in the lie, they say, "I'm sorry I misspoke." Try that in front of a judge and see what it gets you!

CHAPTER SIX

Getting Educated Company Two

The second company I went to work for was located on the Berlin Turnpike in Connecticut. They got their leads through advertising, and every Sunday, they would have their sales flyer in the Sunday newspapers around the state.

The way the sales flyer was set up, it had a picture of two windows, which appeared to be the same window. One window with the bottom sash tilted out and the other with it closed. Also, on the flyer, it said that if you purchased a whole house of windows from us, you would get a free storm door. Every time we went on a lead, the owner would say, "Don't forget to add two hundred dollars for the storm door."

The advertised price was one hundred and twenty-five dollars per window installed. For that price, we got a lot of leads.

Here is how they trained us. We would show up a little late for the appointment. After introducing ourselves, we would apologize for being late because we were the commercial salesmen for the company. We were coming in from out of state, where we just sold a big window job for a new large condo going up in that state, hundreds of windows, and while coming back to Connecticut, my company called me and told me the salesman who was supposed to come here tonight got sick and asked me to fill in for him.

Once we got through showing them the window, a beautiful vinyl replacement window, we would give them the price.

All the time we are showing the homeowners this great window, they are thinking, *"Wow, that window for only one hundred and twenty-five dollars. Let's see, I have fifteen windows that would be eighteen hundred and seventy-five dollars, wow, what a deal."*

Then it was reality time. For the windows, the price came to twelve thousand five hundred dollars. What I really wanted was seven thousand five hundred dollars.

When the homeowners hear this price, they say, "Wait one minute." They go into another room and bring out our flyer that says replacement windows for one hundred and twenty-five dollars each, and tell you what they already calculated in their mind as to how much the windows should be according to our flyer.

When they do that, we would say, "Oh, you want that window." We would ask them to wait a minute while we go get the other window from our car.

This window was aluminum, one step up from a storm window. By the time we showed it to the homeowner, they didn't want it even for one hundred and twenty-five dollars. This is called bait and switch. It was illegal, but this company got away with it for years.

After hearing all the complaints, crying, and disappointment, here is what we would say. "Folks, let me ask you a question. If there were a way, I could greatly reduce the price of these windows, because I know you had your heart set on getting them. Would you be interested?"

They would then ask, "How can you do that? And how much less can you make them?"

I say, "If you folks remember, I wasn't supposed to be the one to come here tonight, because I don't deal with homeowners, only contractors who buy hundreds of windows from me at a time. One time, one of the guys who worked for our company told me his mother wanted new windows, but he didn't think she could afford them.

I had just sold a big job, so to help him out with his mother, I told him to tell his mother that if she ordered the windows right now, I could piggyback her order on
this big order I just sold, giving her the contractor's discount. She ordered the windows, saving a lot of money by getting the contractor's discount."

"So, my question to you folks is if I can piggyback your order on the big commercial order I sold today, saving you a lot of money, would you be interested?"

The homeowners always asked the same question, "How much can we save?" Then I would say I have to call my boss to get approval to piggyback your order first, so let me call him and tell him the situation."

I make the call, and to make the people feel like I'm really fighting for them, I say to the sales manager, "Hi, this is Art, and I'm at the Jones's house. They really like our window and really want to invest in it, but the price is out of their reach. Do you remember how I saved Jim's mother a lot of money by piggybacking her order on a big order I had just sold? Today I sold a thousand windows on a big condo job. Can I piggyback the Jones's order on this order so I can give them the contractor's discount?"

He wouldn't agree to it right away, so I would appear to be fighting to get these people the contractor's discount. After arguing with him for a short time, he would agree.

Now the people felt a little obligated to me because I fought so hard to get them their contractor's discount. Now I tell them that with the contractor's discount, instead of twelve thousand five hundred dollars, the new price would be seven thousand five hundred dollars.

That usually sealed the deal. If need be, I could finance it for them with payments that would fit their budget. When you read it, it sounds like a lot to go through, but it worked! Using this method, I sold a lot of windows for this company.

People really believe they got the contractor's discount, and here's why I know. Once I restarted my business, I got a call from a homeowner for a siding estimate. I noticed they had vinyl replacement windows in their home. I asked them out of curiosity who they bought the windows from. It turned out to be from the second company I worked for. The woman said to me, "And we got a really good deal, the salesman who came here was the one who only sold to contractors, and he got us the contractors' discount, saving us a lot of money."

So, you see, these tactics work!

Some people who enter sales, even with great training, don't succeed because they never learn how to deliver what they are saying properly. I'll get into that later.

Before we go into the next chapter about company three, I want to tell you about a scam the owner of this second company and a tv celebrity pulled.

When I was working for this company, on one of the TV stations was a man who was supposed to be their consumer advocate. If you had a problem with a company in your transaction with them, you would call the station he worked at to tell him the problem you were having with this company, and he would work to resolve it for you.

Every night on TV during the news, he would come on and tell you about someone he helped, and when he finished telling you his name and saying, "I'm in your corner."

It turned out that this man and the man who owned the company number two I was working for at the time were good friends. Every time we would go on a lead, he would tell us to say to the people that he and this TV advocate were good friends and they could even call him to check out the company. He gave us a number for them to call.

One day, I said, "If you and this guy are such good friends, why don't you have him advertise for you?" The owner of this company said he couldn't because he was a consumer advocate, and it would be a conflict of interest. So, this is the scam they came up with.

This so-called advocate for the people decided he was going to help people find a job. On this day, when I went to the place of business for company two, the owner said Hurry up and grab a phone, we are going to be on TV with this advocate, and we'll be getting a lot of calls.

When the advocate came on TV, he proclaimed that our company was hiring people to install windows, showing our people installing windows, siding, and showing our people installing siding, and everything our company did, stating no experience was necessary; the company would train them.

First, I said to myself, What a great ad for the company. Then I realized this company doesn't train installers; they use subcontractors. After the advocate was done, the phones started ringing. People were calling to find out how they could get a job to be trained to do the kind of work we do. We took their names and said someone would get back to them.

After it was over, I said to the owner, "Since when do you train people to do this kind of work? You only use subcontractors. He said, "We're not hiring anyone." It was the only way this advocate could advertise for himself and the other companies he featured in the name of getting people jobs. I'm sure he got paid by these companies to feature them, too.

The station found out about the scam he pulled and fired him. Remember this: when you see a celebrity on TV or hear on the radio advertising for a company, they are doing so because they are getting paid to do so, not because they believe in the company or product. Money talks!

Before I tell you about company three, the next chapter is going to cover advertising and why the companies advertise the way they do.

CHAPTER SEVEN
Advertising Techniques of High-Pressure Companies

Advertising is necessary to make any business successful. If a company didn't advertise, then nobody would know they were in business. There is a saying in the advertising business that goes like this, "I can't understand a man who is in business who won't advertise his business, but when his business fails will advertise his business for sale."

When I had my own business in the early years, we had a bad recession. Where I was living at the time, in that town, there were two car dealerships across the street from one another, one was a Ford dealership, and the other a Buick dealership.

I had done work for both. Business was slow. One day I was talking to the owner of the Ford dealership, and we were discussing the recession and how business was slow.

He said to me, "Art, when times like these come along, I greatly increase my advertising. As you can see, we are very busy. The Buick place across the street stopped advertising, and now they are very slow. "Advertising is the key to getting business," he said. stopped advertising and now they are now very slow. "Advertising is the key to getting business," he said.

So, I immediately placed ads in a couple of different newspapers, and it wasn't long before I had so much work, I had to hire several more men.

Some companies advertise and do it honestly, and some companies use deceit to get you to call them. Their philosophy is, all they want to do is find out who is interested in doing the work on their homes, and once the salesmen get into their homes, they have a good chance of getting your business.

SOME OF THE ADVERTISING TRICKS THEY USE

Have you seen this ad? FOR EVERY 5 WINDOWS YOU BUY, YOU GET ONE FREE!

Order your windows from us today, and installation is free! Window sale, 40% off for three days only

Install vinyl siding on your entire house, and we will give you free replacement windows for your entire house! (There was a company advertising this.)

Get two new entry doors for your house free with new windows for your house.

Install our windows in your entire house and get two free trips to Florida!

The third company I worked for advertised on their flyer, install our window system in your entire house, and we will pay your last year's entire heating bill!

The one I like is "Installation is free." When did anyone ever work for free?

Let's face it, folks, not everybody is going to fall for these come-on ads. Guess what? These companies don't want you folks who know these ads are phony to call them because they can't sell you their product at the high price they want to get, because they know you are too smart to pay it.

These companies only want the people who believe they are going to get something for nothing to call them on the ad. These are the gullible people they gobble up.

There is a saying that has been around for years, and I know you all have heard it, but people still let themselves get scammed. "If it sounds too good to be true, then it probably is."

By the way, the company that advertised that if you sided your entire house with their siding, they would give you free windows for your entire house, ended up eventually changing their ad to "If you don't want the windows, you can deduct

$2,500.00 off the price of the siding job. "Here's the bottom line. Nobody in any business is in business to give away anything for free. It's always figured in the final price you end up paying. And the workers who do the work on your house don't work for free, so the labor isn't free; it's figured in the total price, like everything else you will be getting.

CHAPTER EIGHT
The Power of Advertising

You will understand how powerful this type of advertising is. I am going to tell you this story. I was with the third company, and by this time, I was the top salesman in the company. Later, you will find out what I did to achieve this.

I had a lead in Fairfield County on this night. When new salesmen trainees finished their training, before they could go out by themselves, they had to go out once and observe a seasoned salesperson in action.

On this lead, I had to take one of these individuals with me. Their instructions were to sit and observe and say nothing, no matter what happens.

It turned out the individual I was seeing on this night, along with his wife, was a design building engineer. Now these engineers must know a lot about building materials. So, I had my job cut out for me.

One of the biggest complaints among the salesmen is when they get a lead that turns out to be an engineer. One thing about engineers is they think they know everything about everything, so in most cases, they don't let a window salesman go through his routine; they say, "I already know all there is to know about windows, just tell me the price."

That stops all salesmen in their tracks, that is, unless it's me. I devised a method to not only get them to listen to me but to do anything I asked them that night. I think it's called psychology.

Here is what I would do. The engineers, and I mean all of them, as soon as you sit down at their kitchen table, drop their business card in front of you so you would know who they are. Engineers have very big egos.

It was their way of saying, "I'm smarter than you, so don't try to con me." So, here's what I would do. I pick up the business card and study it really hard, and then I say in a very serious and humble way, "Wow, you're an engineer. This is my lucky day!"

When I said this, he would say, "What do you mean your lucky day?" I would reply, "I just started working for this company. I researched a few companies before I went to work for this one. I felt this company was honest with the best window I ever saw, so that's why I am working for them.

Because you're an engineer, and because you guys know all there is to know about everything, maybe you could help me out?"

He would then ask, "How can I help you out?" I would say, "There are two things you can do to help me out. First, I want you to pay attention to my presentation. Second, I want you to take a good look at my window, the way I show it to you. When I'm done, I would like your professional opinion on my presentation, and your professional opinion on my window.' They all said the same thing," I can do that."

So, for the rest of the night, I had their undivided attention, and they did everything I requested of them, and they were happy to do it. Engineers have a big ego, and I played up to it. Worked every time. Notice I used the word <u>professional,</u> on purpose.

When I asked what they thought of my presentation, they would all say, "Very professional." When I asked them what they thought about my window, they all would say, "Best window I ever saw."

One engineer, when I gave him the prices of the different ways he could do the job, said he wanted to think about it. His wife said, "Honey, you just said it was the best window you ever saw, don't you want to put the best window you ever saw in our house?" She sold that deal for me.

Getting back to the lead with this engineer, where I had to bring a trainee with me. The minute I sat down at the kitchen table, he did what all engineers did: dropped his business card in front of me.

Before I could do my engineering routine, he said, "Wait a minute," and left the room. He came back in the kitchen with a newspaper window ad that said, "For every five windows you buy from us, we give you one free." And put it on the table in front of me.

He said, "Does your company do that? If they don't, then we have nothing to talk about."

The first thing I thought was that this trainee I brought with me tonight is going to learn something he didn't learn in his training. I looked down at the ad and played dumb and said, "Do what?". He said, "Do this". And pointed at the ad. Again, I asked, "Do what?" He picked up the ad, pointed to the part that said he would get free windows, and said, "Do this, right here, it says for every five windows I buy, they are going to give me one free". My next thought was, I couldn't believe a person with the intelligence to be a building design engineer could believe he was going to get free windows. You see, folks, nobody is safe from being hustled.

Let's say, for conversation's sake, his name was John. Here's what I said to him, "John, how many windows do you have in your house?" John said, "18 windows".

I said, "Tell you what I'm going to do, John, tonight you buy just one window from me, and I'll give you the other seventeen windows free." John said, "You wouldn't do that". I said, "John, my man is sitting right there, and your wife is there, two witnesses, if I say I will do it, then you can be sure I will. Again, you buy just one window from me tonight, no questions asked, and I will give you the other seventeen free. Let me help you out. Being a building design engineer, you know the bigger the window, the more it costs, right?

John answered, "Yes, that's right". I said, "So let me help you out. The smallest window in your house is your bathroom window. You buy that window from me tonight, no questions asked, and I will give you the rest of your windows free."

John said, "You're on". I said, "Good, that bathroom window is going to cost you $35,000.00". John said, "Thirty-five thousand dollars, I knew it was a gimmick". When he said that, I picked up the window ad and pointed to the part that said he was going to get free windows and said, "What do you think this is? Do you really think these people are in the business of giving away free windows? You can bet the cost of that window will be covered in the cost of the other windows. Nothing is for free."

The fact is, I walked out of there that night with a contract to put windows in his house. The new man got a good lesson that night.

So, you see, people from all walks of life fall for these phony ads. Let me say it again,

NOTHING IS FOR FREE!

CHAPTER NINE
Getting Educated Company Three

The sales manager at company two was about to quit and go to work for another company where a friend of his was a sales manager. He told me I should go with him because this company was better, and they had a fantastic window.

I went down to Bridgeport, where this company was located, for the interview and got hired right away.

I went through their training, which was the best training I had so far. The night before I was going to go out on my own, they sent me out with what was supposed to be one of their best salesmen.

Unfortunately, the person we were going to see, along with his wife, was an engineer. True to his breed, as soon as this salesman started to speak, he told the salesmen I was with, "I don't want to hear a long story. I already know about windows. All I want to know is how much each window costs?"

This salesman said, "Before I give you a price, I have to measure the windows." But the engineer wasn't having it, and we ended up leaving not being in his house for more than ten minutes.

The next night, I went out on my own lead, didn't sell it. It took me a couple of times before I started getting into the groove with this new window.

Even though I was selling enough windows to make a living, I wasn't satisfied. I didn't mention this yet, but salesmen who sell for these companies only make money if they sell. It's strictly a commission-based situation. The salesman must pay all his expenses, including everything for his car, and he must pay his own taxes. These home improvement salesmen are subcontractors.

After I had been with the company for a while and was doing pretty good, I wasn't satisfied with my sales results and decided to create a new way of selling the windows. This company gave all its salesmen a pitch book. It contained different kinds of information that were supposed to convince homeowners to buy their windows. In the book, it only had some stock pictures of a bay, bow, and garden window. It also had a lot of facts to convince the homeowner that their windows would save them so much money that they would pay for themselves.

I took the best tactics of what I was doing in my own business and the best of what I learned from the three companies and created a new pitch that turned out to be powerful.

So here is what I did. I went to the installation department and told the installation manager to notify me whenever they were going to install a bay, bow, or garden window. Before they installed the windows, so I could take before and after pictures.

One of the salesmen who trained us at the first company I worked for told us how he stopped cancellations. He said, "Before I leave the customer's home after I sold them the windows and got them to sign a contract for them, at the last minute, I would say to them, 'Before I go, folks, there is something I want you to do for me. The last thing my company and I want is for you folks' tonight is to make a decision you are not comfortable with. So, if there is any doubt in your mind, I want you to take all the paperwork you signed tonight for the windows and tear it all up, and we will part good friends, no hard feelings."

He said he would ask them more than once, and they never tore up the contract, and they never cancelled. When I first heard him say that, I said to myself, I could never do that, it's crazy. I didn't believe him and thought something like that would never work.

Well, here I am getting a lot of cancellations and getting very frustrated, so out of desperation, I tried that tactic. Guess what? I never got another cancellation again; it worked! I couldn't believe it. One time, a lady did pick up the paperwork, and it looked like she was going to tear it up, but her husband said, "Honey, we really need the windows." She didn't tear it up.

This company had a sales flyer that they always put in the Sunday papers. The flyer said, "If you install our windows in your entire house, we will pay your last year's heating bill in full.

On one of my leads, when I knocked on the door and an old lady opened the door, and when I introduced myself as the salesman from this company, she said, "Come in, I've got your flyer and I'm going to make you live up to every bit of it."

I now knew I had a sale before I even started. When we went on a lead, one of the first questions we had to ask the homeowners was how much their heating bill was last year, just in case they had our flyer. That way, when we gave them the price of their new windows, we could add the cost of last year's heating bill to it so we could take it off later.

After that, it was clear sailing. One day, the sales manager called me and told me to bring my catalog case down to his office. I asked him why, and he said, "Because you're selling everybody you're talking to, and I want to see what you are doing."

When I got to his office, the first thing he wanted was to see my pitch book. After he looked through it, he said Do your pitch for me.

So, talking to him like I would any homeowner, I did my pitch for him, treating him just like I was trying to sell him the windows.

When I got done, he went to talk to the owner of the company. When he came back, he asked me if I still had the negatives of all the pictures I had in my pitch book. I told him I did. He wanted to know if they could use them. I asked him why.

He said, "We want to change everybody's pitch book to match yours, and then we want you to go to the other offices and this one and re-train everybody on how you sell the windows."

I said I didn't mind them copying my pitch book, but if I spend the time to train all the salesmen in three different locations, what's in it for me? He said, what do you mean?" I said, "I only make money if I'm selling windows. While I'm spending many hours training your salesmen, turning them into super salesmen, I'm not making any money, so what's in it for me?"

He said, "What do you want?" I told him how much money I wanted to train all the salesmen in the company, and it was a lot. But then again, if I were selling instead of training these people, I would be making a lot, too.

He went to tell the owner of the company of my demands. He came back with a counteroffer, which was much lower than what I told him. I said, "The price for me to train all your salesmen is non-negotiable. If you want me to train them, then you have to pay my price." And they did!

The whole concept of this type of sales is to give the homeowner a higher price than they really want, so you can give them an incentive to buy from you that night by saving a lot of money. It's a scam, but other businesses do it too. Here's an example of what I mean.

In my own business, I got a remodeling job for a big furniture store. While we were doing the work on this store, they were going to have a big sale on their furniture. I got to know one of the salespeople there, a lady. In our conversation, she said, "I've only been working here for about a year.

Before we have a sale, the boss raises the prices of everything in the store, so the percentage we take off the sale price ends up giving us the price they wanted anyway." So, you see, other businesses do the same thing, putting a high price tag on their product, so they can offer you a deal, at least you think it is a deal. Deception, just about all businesses use it to get your business.

In the training for this company, I learned things that no other company has taught me. The technique they taught us to give the homeowners the price wasn't verbal. How we did it was to put the estimate on paper, giving the homeowners three different options on how they could do their window job. Also, on this sheet, there would be financing for sixty or eighty months for each method for installing their windows.

Once we were done, we would slide the price sheet between the wife and the husband. Whoever picks up that sheet first is the dominant factor in that family, and if you convince that person to buy your windows, the other will follow. Let me tell you, to my amazement, every time I slid the paper between the wife and the husband, the first one to pick it up was the boss in that family.

We were also taught that when we first enter the house, we must do a warm-up with the people. In the warm-up, there were certain questions we had to ask. Because we didn't know if they had our flyer that says we would pay their last year's heating bill if they put our windows in their entire house, we must ask them how much they paid for their heating bill last year.

If they told us it was a very high amount, then we pretty much knew they had our flyer. So, we would at that time ask them if they had it. In our relaxed conversation with them, we want to find out if they both have a job, are their jobs are secure, for instance, any threat of a layoff, how long they have been living in the house, and if they are planning on selling it anytime soon, and why they think they need new windows.

If you are wondering why we need to know all this, I'll take each one and explain why we need to know the answer to each question.

Question number one: How much was your heating bill last year? We ask this question because whatever they tell us, we must add that amount onto the price of the windows, so we can take it off after making the homeowner think they saved that much money on their windows.

If I must get twelve thousand dollars for your windows, and you tell me your heating bill last year was five thousand dollars, then the price I'm going to give you up front is seventeen thousand dollars. That's always how it works for anything free, no matter what kind of business or product you are buying. All businesses use this tactic one time or another.

Do they both have jobs, and are they secure? Once they tell us there is no chance of a layoff in their jobs, or if it's only the husband working his job is secure, guess what they can't do when we try to get them to sign the contract for the windows? They can't say, "Art, we really like your windows, the price is good, but there is a rumor going around the plant that there might be a layoff. Once I know I'm not getting laid off, I will certainly call you to do the windows." That's an objection we can't get around, but now, because in the beginning I established by his own words that his job was secure, he can't use that excuse not to give me an order for my windows.

The next question is, how long have you been living in this house, and do you plan on staying here? They usually say, "Yes, we are going to retire in this house." Or "We have no plans on selling our house, we like it and the neighborhood too."

Now when I give them the price to escape signing the contract they can't say, and believe me folks people do say this, "Well, we just wanted to see how much the windows would cost because we were thinking we might sell this house and get a smaller or bigger one, but if we decide to stay here, we will call you to buy your windows."

This is another excuse not to sign a contract with you and to get rid of you. But because I established in the beginning from their own words, they plan to never leave that house, they can't use that excuse not to buy my windows. That would be another objection we couldn't get around. In the end, we want the final objection to be about money. That we have no problem getting around, as you will see when you listen to me selling a couple, windows.

The last thing I want to know is why you feel you need new windows? Once you tell me this, then I know what to focus on to get you to buy my windows. I'll give you a couple of examples.

When I pulled up to this lead, what I saw was a pile of firewood stacked high on the side of the driveway. The house was a small two-bedroom ranch. The windows looked pristine. When I got in the house, entering the living room, I saw a huge wood stove in front of the fireplace with the vent pipe installed inside the chimney.

We sat down in the living room, and the first thing I asked them was this. "Folks, I'm a little confused. I see that your house is in pristine condition, your windows are in great shape, and you have a big wood stove that looks like it could heat your whole house. Why do you think you need new windows?

These people were in their seventies, so I knew they weren't worried about a job, and I knew they weren't planning on selling their house at their age, and the great condition the house was in, I couldn't fathom why they would think of spending a lot of money on new windows.

Here's what the lady said to me. "Art, as you see, we keep our house nice and clean. We like to clean our windows at least twice a year. My husband must struggle getting the storm window sashes out, and he must climb up on a step ladder to clean the outside of the windows. I worry about him falling and hurting himself. God forbid he should break a hip; he would never at his age heal right. Now I knew the real reason these people wanted new windows, that's what I focused on and got the sale.

CHAPTER TEN

The Concept of Giving Prices On Paper

Before I went to work with this company, I always gave just one price for whatever job I was trying to sell, even with the two previous companies. Once you give a price like that and the people say they want to think about it, it is very hard to get them to buy whatever it is you are trying to sell them.

By giving them three different ways to do the job with two financing options with each way, it gives you a powerful closing tool, as you will see later.

Salesmen are taught many ways to try and get you to sign a contract with them. It becomes a contest between you and them. I knew a salesman once who wouldn't take no for an answer, and the only way people could get rid of him was to demand he leave their home.

Company three also had a gimmick to get people to sign a contract for the windows. The salesman has tried everything in his toolbox to get you to sign the contract. You've already told him you like the window, the price is ok, and you feel the company is honest, but you just can't afford the windows at this time.

First, when he can't overcome all your reasons why you can't buy his windows now, he would admit defeat. He would ask for a glass of water so he could sit at your table and talk to you while he drags out drinking it. Many times, people will say, "Would you rather have a cup of coffee instead?"

That is music to the salesmen's ears. Now he has time to relax, making you think the sale is over. When he feels the time is right, he would ask the people this question, "Folks, let me ask you a question. If the investment in our window system were lower, would you have given me an order tonight?"

This tactic was usually used on people who didn't have our flyer. This company did other types of advertising to get leads for their salesmen.

The people who think the sale is over usually say yes, they would. What they didn't know was, the salesmen added at least five thousand dollars to the price he really wanted to get. All companies have a price they must charge to be able to do the job and make a big profit. This is called the par price. The salesmen must bring the job in at least at the par price. When he brings the job in at the par price, he gets ten percent commission on the gross sale. On a twenty-thousand-dollar par price job, his commission would be two thousand dollars.

If he brings a par price job with two thousand dollars over the par price, then he would get ten percent of the par price and fifty percent of the overage.

Back to our salesmen having coffee with the potential customers. They just told him that if the windows were lower, where they could more easily afford them, they would have given him the order.

He says, "You know, folks, advertising in any business is a big expense. The best advertising is word of mouth from previous satisfied customers, wouldn't you agree?" People always agree. Then the salesmen would say, "Let me run this by you to see if it's something you might be interested in. It might save you a lot of money. The manufacturer of our window system has given our company an allotment of money to get customers to help us sell our window. The state is broken down into sections.

There is only one allotment to a section, first-come, first-served.
If there is still an allotment in this area and you could get it, and it saved you enough money on the total investment of our window system, would you be interested?"
The question the people always asked was, "How much are we talking about?"
The salesmen would say, "I don't know, but I do know they are usually a lot. If you're interested in this program, I would have to call the office to see if one is still available in this area, but first, I must get your agreement on a few requirements to get this allotment."
"To get this allotment, you must agree to let us keep a job sign in your front yard for sixty days, you must agree to tell people about us and our window, recommending our company to them, and in one year, we need a letter from you telling us how much you like or dislike our window, just be honest." They always agree.
Before I call my sales manager, I say to them, "Folks, please don't embarrass me when I make this call. All I ask is that you give me an answer whether or not you want to take advantage of this program. I can take a yes as well as a no." They say they will.
I made the call. When the sales manager answers, I tell him who I am and whose house I'm at, and tell him the people I'm with like our windows and want to have us install them in their home, but the investment is a little high for them.
The reason I'm calling is to find out if there is still an advertising allotment for this area. He tells me there is one, but other salesmen in the area called him and told him he had people deciding whether or not they wanted in on the program, and would I hold it for him. That's when I would act like I got mad and tell him that at the meeting you told us this program was supposed to go to first-come come first-served serve and if that other salesman's people haven't decided yet, then these people are ready to decide.

Then he would ask me if these people agreed to all the terms so they could get the allotment. I would say yes, they agreed to all the terms to get on the program. Then he would ask me if they were ready to decide right now. And I would say yes, they are ready to decide right now.
This is all an act. It is intended to make the homeowner feel like I'm really fighting for them. He doesn't know how much money I have to play with, so he says ok, give it to them. I respond, "They can have it, that's great; how much is the allotment?" I say, "Wow, that much", just a minute. I turn to the people and say, "There is another salesman in the area, and he is trying to get his people to accept the allotment, but as of right now, it's available!
If you want to take advantage of it, my boss needs to know right now, so nobody else can get it. The allotment is five thousand dollars. Folks, when did you ever make five thousand dollars so fast?"
By this time, the people are excited and involved and usually agree, and I have a sale. I told the sales manager they want the allotment, so the other guy is out of luck, and I hung up and did the paperwork.
After I developed my new system of selling, I usually didn't have to use this tool to get them to sign; I usually had them signed up at the high price. I would either tell them about the program after they signed the contract, not wanting to charge them too much over the par price, or offer them a new prime door or two to get them on the program.
For instance. Let's say I sold the window job at a high price. Once I had the contract signed, then had my water or coffee to get the people relaxed, and after I asked them to tear up the paperwork, I would put everything away and get up to leave.
I'm still at the high price, $5000.00 over the par price. Instead of giving them money to get into our advertising program, when I open the door to leave, I put my catalog case down and say, "You know folks, it's a shame you will have all new energy-efficient windows in your home, but these old doors are energy losers too.

How would you like two new energy-efficient doors also? They would always say, "Oh no, we don't want to spend any more money tonight."
I would say, "I'm not asking you to spend any more money. How about if there was a way I could give you two new doors, the style of your choice, for free!"
They would say, "For free, how can you do that?
That's when I would explain the advertising program to them, and like I said.
Instead of taking money off, I would give them the doors, giving me a bigger payday. And they always believed they got the doors for nothing. Even if a friend said I used that as a gimmick to get them to buy the windows, they would say, "No, he didn't, he gave us the doors after we signed the contract to buy the windows."
There is no chance of a cancellation this time.
When you read the chapter titled "The Sales Pitch," you will see what I did to get people to sign at the high price.
This company didn't allow anyone to come in thousands of dollars over the par price. Up to five hundred dollars was ok, no more than that.
Sometimes after trying everything, he couldn't get the deal, and he still had enough money left above the par price. Many salesmen would use this tactic. And believe me, this tactic would make many people feel sorry for the salesmen and end up giving them the contract to do their job. In a last-ditch effort to get the sale, some salesmen would say, "Folks, I work for commissions only. But you folks seem like nice people, and I know you really like my windows and need them badly, so to help you out, I will give up my commission, which will reduce your investment for the windows by another two thousand dollars, which would have been my commission. If I do that, would that make the deal a better one for you?"

CHAPTER ELEVEN

Sales Training by Me

As I mentioned previously, Company Three paid me to train all their salesmen at their three locations. Even after I started my own business again, over the years, other companies hired me to do the same.

I would always start a new training class with this little story. "A young priest one day went to see Father Superior. When he got to Father Superior's office, he asked the young priest how he could help him. The young priest said, "Father Superior, is it ok if I smoke while I pray?"

Father Superior responded, "Absolutely not! That would be a sacrilege against God. It's disgraceful you would even think of such a thing. Leave my office immediately!"

The young priest left Father Superior's office with his head hung down, with a sad look on his face. On his way back to his room, an older priest saw him and saw that he was in a sad state and said, "What's the matter, my son, you look very sad? Did something bad happen to you?"

The young priest said, "I just asked Father Superior if it was ok if I could smoke while I pray? He got mad and yelled at me and made me leave his office, refusing my request. The only reason I asked him is because I always see you smoking while you pray."

the old priest," I see, when I went to see Father Superior, I asked him if it was ok if I prayed while I smoked."

So, you see, it's not what you say that counts, it's how you say it.

Next, I would go into detail on how to begin their evening with the potential customers from the very beginning to the end.

When you are approaching the person's home, observe what is in their yard: a boat, a garden, Etc. You can use these to create a conversation in the warm-up to relax the customer by asking the questions about what you saw, getting them to open up and talk about themselves.

Try to do your warm-up in their living room. This is the room people go to relax. In the warmup, you must get answers to certain questions to stop objections they can use when you are trying to get them to sign the contract. How did you learn about our company? Do they have a good and secure job? How long have they lived in this house, and do they plan on staying there? Why do they feel they need new windows (or whatever you are trying to sell them)? Also, you want to know who might have been there before you, so ask them if they have been looking for new windows before you came along.

At all times, be relaxed, talk to these people like you would talk to a friend. If you show these people, you are nervous, unsure of yourself, and can't answer their questions accurately, then you will lose them and not make a sale.

Once the warm-up is finished, you must get them into the kitchen. It is a proven fact that people make most of their important decisions in their kitchen.

Once in the kitchen, you can start your presentation and window demonstration. Make sure you get the homeowners involved in operating your window sample.

Always ask questions to get positive feedback from the homeowners. The smaller yeses you get now, the easier it will be to get the big yes at the end of the night when you are trying to get them to sign the contract.

When measuring the windows, let the homeowner assist you by writing down the measurements. The more you get them involved, the easier it will be to sell them your product.

Never badmouth your competition. This turns people off, making them think you are doing that to get them to believe you are the only honest person in the business.

There is a saying in sales, "If I tell you something you can doubt it, but if I get you to tell me, then there is no doubt in your mind." The trick to successful sales is not by telling and telling, but by leading the potential customer by constantly asking questions that give you the answer you want. Selling by leading questions, not telling.

Once you slide the price between the wife and husband, no matter what they say, keep your mouth shut. He who speaks first loses. Let them say their piece. Use the closing techniques we teach you here. (For those of you who don't know what closing means in sales, it means getting the potential customer to sign on the dotted line, giving you the order for whatever you are selling.)

After the sale is over, don't rush out on them. You must bring them out of the clouds, so to speak. It's been a high-tension evening. You must get them relaxed before leaving. If you don't, you will get a cancellation.

Ask for a glass of water before you go. Most of the time, they will ask you if you want a cup of coffee. Take it and take your time drinking it, having a relaxing conversation with them. Now you are friends. This will make the sale more solid. Remember, to be successful in sales, you must perfect your approach, your speaking skills, be organized, knowledgeable about what you are selling, and come across as sincere, not a phony. Act calm and stay calm. If you appear nervous or not knowledgeable about what you are selling, then you will lose the deal. When you finally give the homeowners the price for what you are selling, and their response is, "Are you kidding me? No window is worth that much money!" That response means you didn't do your job right and might as well pack up and go home, because nothing you do after that is going to get you the sale.

I want to talk about keeping your mouth shut once you give the price. No matter what the people say after giving them the price, unless they tell you you're crazy, keep your mouth shut and do not defend your price.

Let them rant and rave, which means you made them really want your product. Example: When I slid the price sheet between the wife and the husband, the wife picked it up right away, looked at it, and said, "Thank you for coming." The husband said, "What's wrong, honey?" She said, "Just look at these prices." He took the sheet from her and looked at the prices, put the sheet down, walked over to the window sample, and said, "But honey, look at this window." He began demonstrating the window to his wife, and she finally agreed to give me the order because he really wanted them.

If I had said anything at all when she said thank you for coming, I would have lost the sale. So, I let the husband sell her the windows, keeping my mouth shut.

This lead I was on was when I was in my own business. This is another example of not defending your price. This lead was for a siding job. First, I designed my aluminum trim in a way we didn't have to use J-channel around windows, doors, or most any other places J channels were used. This took a lot more aluminum to make and longer to install. Also, I used one-inch-thick insulation with aluminum foil on both sides, not the thin insulation most everybody else was using.

For the gutters, I always used the Alcoa floating gutter system, the best gutter system you could ever put on your house if done properly. Don't fall for that new gutter that has its own cover as part of the gutter with the opening towards the front. They claim nothing can go into that gutter, but leaves and other debris do, in fact, enter that gutter. You can have quality covers put on the Alcoa gutter system that will, in fact, keep your gutters clean.

When I gave this homeowner the price for the siding job, he immediately said to me, "Art, we really like the way you are going to do the job, but you are about four thousand higher than the other guy we had here.

I knew the other guy and how he did his siding jobs, so here is what I said. "I can meet his price." He said, "You can?" I said, "And this is how I will do that. Instead of putting a heavy-duty Alcoa floating gutter system on your house that I guarantee the labor for fifteen years, I will put the same thinner gutter with spikes and farrows on your house, like he is going to do instead, with only a one-year guarantee, because gutters installed like this end up falling off the house eventually anyway.

Under the siding I will use the thin insulation like he is going to use instead of the one-inch insulation with foil on both sides, and instead of the fancy trim I showed you that takes a lot more material and time to custom make for your house and install, I will cover your trim with the standard way using j-channel like he is going to do."

After I said all that, the homeowner said, "But Art, we want it done the way you showed us." I said, "If you want it done the way I showed you, then it will have to be the investment I gave you." He looked at his wife and said, well, I guess if we want it, the Art wants to do it, I guess we have to pay his price." And they did.

This is a true story, and I only tell you this so you can see that defending your price doesn't work, but showing the people, like I did, the difference in the value of the two jobs made the deal happen.

CHAPTER TWELVE
The Sales Appointment

When reading this chapter, you will experience me actually performing a sales meeting with a married couple; for this, we will call them John and Mary. When you see print italics underlined, that would be me thinking to myself. It will start as I approach the house.

There's the house. What a nice garden these people have. House looks in good shape. Two nice cars in the driveway. I'll pull up in the driveway.

I get out of my car, gather up my sample and catalog case. (I use a catalog case because a brief case is too clumsy, and a catalog case can sit on the floor next to me, making it easy to get the things I need to perform the sale.)

I take my window sample and catalog case with me to the front door and ring the bell. A lady answers the door. I introduce myself, telling her my name and the company I represent. She invites me in.

Her husband comes out and introduces himself, and we shake hands and sit in the living room.

I start the conversation by saying, "This is a nice house you have here. Have you been living here for very long?"

John replies, "Yes, we've been living here for 20 years, we like this area and this neighborhood too."

I say, "So I guess you folks aren't planning on moving anytime soon then?"

Mary says, "Oh no, we plan on retiring in this house, we would never move." I then ask, "So John, if you don't mind me asking, what kind of work do you do?"

John says, "I'm a jet engine mechanic for Pratt and Whitney."

I say, "That sounds interesting, John. Are you guys busy?"

John says, "Oh yes, the place is booming, we can't keep up with the orders for our engines."

I say, "Great, how about you, Mary, do you work?"

Mary says, "Why yes, I do, I'm a legal aid at a law firm."

I say, "I guess I don't have to ask you if you folks are busy; law firms are always busy."

Mary responds, "Yes, they are."

My next question is, "How did you folks come to call our company?"

John says, "We didn't call your company; we saw your company's display at the home show and signed up there to have someone come out to give us an estimate for windows for our house."

I reply, "Great! So, John, Mary, what made you decide you wanted new windows?"

John replied, "First of all, our windows are old and drafty even with the storm windows. Second, they are very hard to clean. And third, we want to save some money on the cost of heating our house in the wintertime."

That's when I ask if we could move to the kitchen because it would be easier for me to show them the information in my book on the table and demonstrate my window for them. John says Sure, and we move to the kitchen.

I open my window sample, and when it opens, the case ends up being the stand for the window. Next, I go to my pitch book.

In the pitch book, I have many facts that show how my window will save them money on their energy cost for heating and air-conditioning their home.

While showing them the facts in my pitch book, I ask them questions to get feedback from them, which lets me know their level of interest and believability in what I'm saying.

Then I show them before and after pictures of all the different styles of windows we sell and install, which gets a positive response from them both. Now I move to the window sample. I demonstrate how to tilt it in for cleaning. I have Mary practice doing it, then John. When that phase of my presentation is done, I ask them if there are any questions. I answer any questions they have. The usual one is this.

Mary says, "The windows look great. How much are they?"

I say, "Well, Mary, first we have to measure them, and John, if you will help me by writing down the measurements for me, it will expedite the process."

John says, "I'd be happy to." We measure the windows.

After we measure the windows, I figure out the three different ways we can install the windows. Triple glass with aluminum casings covering the wood trim on the outside of the old frames. (Replacement windows fit in where the sash of the old window was, leaving the wood frame on the outside the same, with the sixty and eighty-month payment schedule on the side. Next, I figure the price for the windows with double glass and casing covers, with the sixty and eighty-month payment schedules, and lastly the window with double glass, no casing covers, and the payment schedules.

Once I get it all down on the estimate sheet, I tell John and Mary, "John, Mary, there are three different options you have to order your windows in the way you prefer to have them installed in your home."

I then slide the price sheet in between them, and John picks it up. John and Mary, both look it over for a couple of minutes and put it back on the table, and John says, "Art, the windows are a little more than we expected, so we will have to think about it and get back to you."

At this point, there are a lot of ways I could try to close them, but in a lot of cases, it only ends up in them giving me more excuses and it pushes them away from me so I relent and say, “I don’t blame you John, it’s a big investment and you should take your time with it.”

I feel confident in what I did now because all through the presentation, I got all the positive answers I needed from them to know I did my job right and created the desire in them to want my windows. Remember, I said earlier in the book, when the homeowner says the price is too high, you’re about to find out why. I take the price sheet and put it in my catalog case, close it up, and close my window sample, making it look like I’m getting ready to leave.

(What they don’t realize is I’m just getting started.)

When it looks like I’m finally ready to leave, I ask Mary if I could get a glass of water before I leave. She says, “Sure, Art, would you rather have a cup of coffee instead?”

I say, “I’d love a cup of coffee.”

Everybody is relaxed now; the pressure of the sale is over. We are just some friends having a conversation. After a while, when I think the time is right, I say, “Well, John and Mary, I guess I'd better get going now, but before I go, I wonder if you could do me a favor?”

Mary says, “What can we do for you, Art?”

I open my catalog case and take out the price sheet and place it back in between them and say, “Just for conversation's sake, hypothetically speaking, if you had given me an order for the windows tonight, which one would it have been?”

John and Mary pick up the estimate sheet, discuss the three different options, and, as usual, they pick the middle-priced one and tell me that’s the one they would have gone with.

Once they make their choice, I then ask, “And which one of these monthly investments would you have picked?” (Did you notice I didn’t say payments.

When you pay for something, you only get what you pay for, nothing more, but when you invest, you get back from your investment more than what you paid for.) Now we find out the real reason they didn't give me the order tonight.

John says, "That's the problem, Art, both monthly payments are out of our reach. We couldn't fit either one into our budget."

I then ask, "John, how much more do you think you could have managed in your budget?"

They both discuss it and finally John says, "Well, we could fit about ninety more dollars in our budget without too much trouble."

You must understand that these people think this is all hypothetical and feel comfortable answering all these questions.

I ask, "So you're telling me if I could have put windows in your house tonight for ninety dollars a month, you would have given me the order?" Thinking it's over, they say yes.

I then say, "Let me see what I can do," and I go back into my rate book. I say, "John, Mary, I can put the windows in your house tonight for eighty-five dollars a month, five dollars less than what you told me you could afford. Do we have a deal?"

How are they going to say no now? They just told me that for ninety dollars a month, they would have given me the order. Nobody ever said no to me on that one, and I still had them at the high price. So, I would end up giving them a couple of new prime energy-efficient doors for free if they would get on our advertising program.

After doing all the paperwork and getting them to sign everything, I still asked them to tear up all the paperwork if they didn't feel like they made the right decision that night. Getting what they thought was two new free doors also helped keep them from canceling on me.

I sold a lot of windows using this method, and back in my business, I used it to sell everything we did. The only difference, in my own business, the price I gave was the real price I had to get to do the job.

This method also works well when the people you are trying to sell keep telling you they can't make a decision right away because they have to think about it.

So, when I finally got them to hypothetically choose the one, they would have gone with, guess what? They said they couldn't make a decision that night, but now I've gotten them discussing which way they would have gone, and they end up making the decision they claimed they couldn't make tonight."

The one situation I like the most is when people tell me they are so deep in debt they couldn't squeeze another penny into their budget. I'm going to tell you about one of the situations I came across for this very reason.

The third company I worked for got its leads in several different ways. Newspaper advertising, radio advertising, putting their flyers in several different Sunday newspapers, and they also made cold phone calls soliciting people for appointments.

The appointment I had that night was a cold call phone lead. When I got to the house and rang the bell, the man of the house answered the door. After I introduced myself, he said, "When my wife told me she agreed to have you come here to show us windows and tell us how much they cost, I told her I wasn't interested in windows right now and couldn't believe she made the appointment without asking me first. I'm sorry you came all this way, but I'm really not interested in any of this right now."

I said, "No problem, I understand, but can I ask you a question?" He said Sure."

I said, "Many experts predict that in the not-too-distant future, the cost of gas for our cars and the cost of home heating oil is going to go to around five dollars a gallon. When that happens, do you think you would then be interested in energy-efficient windows that would save you a lot of money on your heating bill?"

He said, "Sure, I would be if the price went that high."

I said, "I traveled a long way to get here. I'm going to get paid whether or not I show you our window system tonight. (A lie) Now that I'm here, why don't you let me show you the window, let you know what your investment would be, and later down the road, when the prices of heating oil get high, maybe then you'll want the windows.

I'll leave my card, and maybe you'll call me then." He agreed. By the time I got through with my presentation and giving them the price and doing the hypothetical with them and asking them how much more they could fit into their budget he said to me, "I know you couldn't see it from the road, but we just bought a solar system that's on our roof for over eight thousand dollars and I'm making payment on it. That, along with all the other bills we have, we couldn't possibly afford another thing right now.

He also said, "I really wish I could get your windows, but I just can't afford them right now."

That was music to my ears. Here's what I did now. I said, "Let's forget for now I'm a window salesman trying to get you to invest in my window system. Look at me as your financial advisor. Bring me all your bills that you must pay. What I'm going to do is figure out how much it would take to pay all your bills off and put the window in your house.

What you will end up with is paying out a lot less every month with all your bills gone, new windows, and a lot of extra money in your pocket."

That's just what I did and got another window sale. This works with people in debt from financing too much because they are chronic finance buyers. Finance buyers when they see something they want will finance it to get it now. Regardless of the added cost of interest. As long as they can fit it into their budget, they're buying. What I just did is called the consolidation close, the most effective close in-home improvement sales. Works every time. I never lost a consolidation close, only if they didn't get approved by the loan company.

When most of these companies train their salesman, they teach them to keep asking certain questions throughout the entire presentation. For instance, when they tell you about their company, they work to make you think they are the best company in the business. They ask you, "Is this the kind of company you want to do business with?" Most people would say yes. (Another small yes leading you to the big one.)

When they demonstrate their window and give you the details of it, they ask you, "Is this the kind of window you would like in your house?" Of course, people most of the time say yes. (Another small yes leading you to the big one.)

Every time they make a point, they will ask you, "Does that make sense to you?" Every time you answer yes, they are getting you closer to buying whatever they are selling.

Everything a home improvement salesman does from the minute you open your door until he gets you to sign a contract is to separate you from your money. He has to do it then and now because the company won't let him have another chance at you, and if he wants to get paid, he has to sell you and you to sign a contract.

In his training, he is told, "When you go into the prospective customer's house to sell them whatever it is you are trying to sell them, there will be a sale made there, either you will make it, or the homeowner will make it." (Meaning, if you didn't sell the homeowner that night, the homeowner made the sale, and you failed.)

CHAPTER THIRTEEN
Different Types of Cons

I got a cold call from a guy who claimed to be a financial planner, and he had this great investment that was going to make me rich if I invested my money with him. Out of curiosity, I listened to his pitch and answered his questions, giving him the feeling he had a big fish on the line, and when he thought he had me hooked on his scam, he said, "So how much money do you have today to risk on this investment?"

I said, "None, I only invest in a sure thing." He hung up on me.

When you get a cold call from someone telling you he or she has a fantastic deal to offer you that will make you rich or live many years longer or anything else that seems too good to be true, hang up; it's a scam.

These people are like trout fishermen. The trout fisherman throws the fake fly in the river to hook the fish. These people are like that trout fisherman. The unbelievable offer they are telling you about is like that fake fly they are using to hook you too. Be aware of the trout fisherman who tries to sell you that miracle.

These con artists are good. The third company I worked for before starting up my business had a sales manager who was one of the best salesmen I ever met. We became good friends. He left the company before I did, and after I started up my business, he hired me to do a major do-over of his house.

One day, he called me up and said, "Arthur, I got in on a great deal today, you have to get in on it, you will make a ton of money."

I asked him what it was about. He said, "A guy who is a financial planner called me and told me about a new stock that came cut. He said If I invest in it, within six months the stock would split, and in a year, they will split again, making me thousands of dollars.
So, I invested five thousand dollars with him." I said to my friend, "As good a salesman as you are, I can't believe you fell for that scam. You just got scammed out of five thousand dollars. You will never see that money again.
He was adamant, "Art, this guy is legitimate, believe me, this is real."
About six months later, I asked my friend how his five-thousand-dollar investment was going. He said, "There have been some delays in the stock splitting, but the financial planner says it's going to happen soon."
Bottom line, my friend not only didn't get rich from this investment, but he also never got his five thousand dollars back either.
I'm telling you these stories because the purpose of this book is to educate you and keep you from getting scammed. If you have ever had dealings with the IRS, as I have, then you know they like to talk to you like you are something stuck to the bottom of their shoe. They have a snotty attitude when they talk to you.
One day, when I was home, my phone rang. It was a woman who gave me what was supposed to be her name and said she was from the IRS, with that typical snotty attitude sound in her voice. She said, "The reason I'm calling is because we want to know why you haven't paid your income tax yet." I said, "I did pay my income tax."
She said, "But you didn't pay your income tax on the three hundred and fifty thousand dollars you won on the European lottery."
I said, "How could I have won money from a lottery I never played?"
She said, "I don't know, but we have the check right here made out in your name for three hundred and fifty thousand dollars."
I said, "You do? Well, send me the check along with the tax bill, and I will be more than happy to pay for it."

When I said that, she said, "I can't send you the check until you pay the government entry fee into our country of three thousand dollars. Once you send us that money, we can send you the check."

So, I said, "Tell you what you need to do. On an IRS letterhead, send me all the details and your requirements for me to receive this check you claim I won. When I get this letter from you, I will send it to my brother who works for the IRS After he checks this out and tells me this is all legitimate, I will be more than happy to send you three thousand dollars to get a three hundred- and fifty-thousand-dollar check."

She said, "I'll do just that." She hung up, and I never heard from her again. Why have I told you about these scams? To let you know how crazy and unbelievable some of these scams are, yet many people fall for them, even so-called super smart people.

I especially like emails I've gotten many times. It is someone who is supposed to be from another country. His father or someone in his family left him millions of dollars, and the only way he can get the money out of his country is if he can transfer it into my account. If I do this for him, he will pay me a big commission when he comes to the USA to get it. All I have to do is give him my account information so he can transfer the money.

Once he has that information, he will be able to clean out my account.

As crazy as all the scams seem, many people do fall for them.

No matter who calls you, no matter what they say, no matter how much they try to scare you into acting immediately, before you do anything, verify what they are telling you to make sure it is the truth and not a scam.

CHAPTER FOURTEEN

Finding A Reliable Contractor

Getting back to the home improvement business, remember this. When someone is trying to sell you any kind of work to be performed in our home, once they give you a price and you tell them, for whatever reason, you are not going to sign up with them at that time, and all of a sudden, they come up with some reason why they can reduce the price, show them the door.

An honest contractor doesn't have to inflate his price to trick you into signing up with him.

When you deal with a company who have salesmen, subcontractors, office personnel, plus the cost of their office space, in essence, you end up paying for all that and their huge profit too.

How to find a reliable, honest contractor.

Call the Better Business Bureau.

Check with your local building inspector to see if he knows the contractors you are thinking about doing business with, and if he has had any trouble with them. Check with your state's consumer protection department and see if they have received any complaints about the contractors you are thinking about getting estimates from.

No matter what any of them say to get you to sign a contract on their first visit, get three estimates before you decide on who you hire.

Before you call any contractors, check them out first, so the only ones you end up getting estimates from are already cleared in the manner I told you. When you finally sign up with a contractor, do not give him any money up front. If he insists, tell him that when he delivers all the material to your home to do the job, and he starts the job, and with the guarantee that they will come every day until they are done, you will give him fifty percent of the job. If he is getting you financing for your job, then you don't have this worry. He can't get paid by the finance company until you sign off on the job.

When you finally find three contractors who check out the way you want and get their estimates, I want you to understand this one fact. The lowest price isn't always the cheapest price. Especially if the work you have done is on the outside of your house. Mother Natre is a tough old lady; if you cut corners on her, she will make you pay.

When getting your estimates, make sure you understand how the job is going to be accomplished and what type of material is going to be used. Make sure you get at least three references from past jobs they did, so you can check out their work and see if these people are satisfied with the job that was done for them.

If you don't know what to look for, then you can't see what went into a siding job or the quality of the window you are getting. People I've given an estimate to would say, "One window is as good as the next; they are all basically the same." They couldn't be more wrong about that.

There are some cheaply made windows with poor quality parts that don't last very long.

Remember this, and it amazes me people don't. You think nothing about buying a car for thousands of dollars and losing thousands of dollars as the years pass due to depreciation, but want to go cheap on your house, which doesn't depreciate, instead continues to increase in value. And do the improvements on your house also increase in value?

Remember, today is the cheapest day you will be able to have work done on your house, because tomorrow the price will be higher. Prices are always going up. When I first started in this business, I paid a little over seven dollars for three bundles of roof shingles; now the cost is around fifty dollars for one bundle. You want to get the job done on your home for a fair price for what you are getting, which doesn't necessarily mean the cheapest price.

If you get a quality job done on your house, it will not only pay you back in energy savings and maintenance expenses, but it will also increase the value of your home.

A word of caution. If the contractor gets you financing for the work he is going to do for you, you probably will end up with a lien on your home for that money until you pay the loan off. Most contractors don't tell the people they are dealing with about the lean because they don't want to scare them away from signing the contract.

This turned out to be a big problem in my state some years ago. When you sign this kind of loan, there is supposed to be a notary present to notarize that you actually signed for the loan because it would become a lien on your home.

Because the contractors didn't want you to know this, they had someone they paid to notarize the paperwork after they left your presents, not letting you know about the lien. That is illegal!

The problem arose when people tried to get a second mortgage on their homes for whatever reason, and they were refused because they had a lien on their home from the home improvement loan they had.

The attorney general got involved because these loans were obtained illegally, because they were not notarized legally, and I believe the loans were made null and void, releasing the homeowners from the loans.

The home improvement contractors aren't the only ones who use these types of tactics to separate you from your money, so be careful when making large purchases. Take your time when making big money purchases. If you do, you will avoid being ripped off.

CHAPTER FIFTEEN
Objection Handling Used Against You

Your Words Can Be Turned into Weapons. You think you're being cautious. You raise a concern. You ask a question. You say, "I'm not sure." But instead of slowing down the sale, your hesitation becomes fuel for the closer.

That's not an accident. It's training. Salespeople are taught to expect objections — and to use them as stepping stones, not roadblocks. They don't fear your resistance. They welcome it. Because every objection you raise gives them a map of your mind.

The Psychology Behind Objection Handling Objection handling is the art of turning "no" into "yes" — not by solving your problem, but by reframing it.

Here's how it works:

You say: "It's too expensive."
They say: "Compared to what? Can you really put a price on peace of mind?"
You say: "I need to think about it."
They say: "Of course — but what specifically are you unsure about?"
You say: "I want to get other quotes."
They say: "That's smart. Just make sure you're comparing apples to apples — and remember, the best companies book up fast."

Each response is designed to keep you engaged, off-balance, and moving toward a decision — their decision.

The Real Goal: Control the Frame

In sales, the "frame" is the lens through which the conversation is viewed. If they control the frame, they control the outcome.

If they frame your hesitation as fear, they can "reassure" you.

If they frame your delay as risk, they can "protect" you.

If they frame your logic as overthinking, they can "simplify" things for you.

The more you talk, the more they learn. And the more they learn, the more they can steer.

The Flip: How to Regain Control

Here's how to turn the tables:

1. **Don't Reveal Your Objections Immediately**
   Hold your cards. Ask questions instead.
2. "What's your cancellation policy?"
   "Can I see that in writing?"
   "What happens if I change my mind?"
3. **Use Neutral Language**
   Avoid emotional cues they can latch onto. Instead of "I'm worried," say "I need more information."
   Instead of "I'm not sure," say "I'll review this on my own time.
4. **Break the Frame**
   Interrupt the script.
5. "I don't make decisions under pressure."
   "I'll reach out if I'm interested."
   "I'm not here to be closed — I'm here to be informed."
6. **Watch for Pattern Interrupts**
   If they suddenly shift tone, mirror your language, or start "agreeing" with everything you say — they're adapting. That's not empathy. That's strategy.

AI and Automated Objection Handling

Now imagine this process — but powered by AI.

Chatbots that respond instantly to your hesitation.

Algorithms that test different rebuttals in real time.

Voice assistants that learn your objections and refine their pitch.

This isn't the future. It's already happening

Digital systems are being trained to handle objections just like human closers — but faster, more persistently, and without fatigue.

Your Defense Strategy

Silence is power: You don't owe an explanation.

Delay is control: Time breaks momentum.

Documentation is leverage: Get everything in writing.

Exit is freedom: You can always walk away if you aren't in your home.

CHAPTER SIXTEEN
The Contractor's Play Book
What They Hope You Don't Know

Not All Contractors Are Crooks — But All Have a Playbook.
Let's get one thing straight: there are honest contractors out there. Professionals who care, who deliver, who stand by their work. But even the good ones use tactics, and the bad ones use those same tactics to manipulate, mislead, and maximize profit at your expense.

This chapter isn't about paranoia. It's about preparation.

You'll learn how contractors think, how they sell, and how they structure deals to protect themselves — not you.

The Psychology of the Pitch

Contractors don't just sell labor. They sell confidence. Urgency. Authority. They know that most homeowners feel overwhelmed, underinformed, and desperate to "just get it done. "So, they lean into that.

They speak in absolutes: "This is the only way to fix it."

They use fear: "If you wait, the damage will spread."

They create urgency: "We're booking up fast."

They simplify: "Don't worry about the details — we'll handle everything."

These aren't just words. They're tools. And they're used to steer you toward a signature.

The Contract Trap
Here's what most homeowners miss:
Vague language: "Install premium materials" — but what does "premium" mean?
Hidden exclusions: "Labor included" — except for demolition, cleanup, or permits.
Change order clauses: One small tweak, and the price jumps.
No cancellation terms: Once you sign, you're locked in.
Contracts are written to protect the contractor. Your job is to rewrite the rules — or walk away.
The "Free Estimate" Illusion
Free estimates aren't free. They're strategic.
They get a foot in the door.
They create a sense of obligation.
They allow the contractor to "diagnose" problems — real or exaggerated.
Once they've invested time, they expect you to reciprocate. That's not generosity. That's leverage.
The Upsell Game
Watch for these phrases
"While we're here…"
"It makes sense to do this now…"
"You don't want to patch over a deeper issue
Upsells aren't always scams — but they're often pre-planned. The initial quote is bait. The real profit comes later
The Digital Contractor
Today's contractors use tech to amplify their pitch:
AI-generated proposals that look professional but hide vague terms.
Automated scheduling tools that create false urgency.
Review platforms are flooded with fake testimonials.

Even good contractors may unknowingly use tools that manipulate, because the software was designed to close deals, not protect consumers.

Your Defense Strategy

Get multiple quotes — and compare them line by line.
Demand specifics — materials, brands, timelines.
Read every clause — especially cancellation and change orders.
Never sign on the spot — pressure is a red flag.
Verify credentials — licenses, insurance, references.

CHAPTER SEVENTEEN
Building Your Personal Defense System

Knowledge Is Not Enough — You Need a System. You've seen the tactics. You've felt the pressure. You've spotted the scams. But awareness alone won't protect you. You need a system — a repeatable, reliable way to evaluate any offer, any pitch, any contract. This chapter gives you that system.

Whether you're dealing with a contractor, a salesperson, a website, or an AI-powered chatbot, this framework will help you slow down, think clearly, and make decisions that serve you — not them.

The Five Shields of Consumer Defense

1. Pause the Process

If they're rushing you, they're controlling you.

Take a breath. Step away. Sleep on it. Time is your ally.

2. Ask Strategic Questions

"Can I see that in writing?"

"What's your cancellation policy?"

"What happens if I change my mind?"

"Who else has used this service, and can I speak to them?

3. Demand Specifics

Vague language is a red flag.

"What brand of materials?"

"What's included — and what's not?"

"What's the timeline, and what happens if it slips?"

4. Verify Independently

Don't trust — verify.

Don't engage with chatbots that push urgency or emotion.

Read privacy policies and terms — or use tools that summarize them.

AI systems are trained to close deals. You must be trained to resist manipulation — even when it's wrapped in convenience.

Bonus: The Tactical Defense Card

1. Pause — No rush.
2. Ask — Strategic questions.
3. Verify — Independently.
4. Document — Everything.
5. Decide — On your terms.

CHAPTER EIGHTEEN

The Rise of AI in Consumer Transactions

The Salesperson You Can't See — But Can't Escape. You're used to dealing with humans — contractors, closers, reps. But now, the pitch is coming from machines. Algorithms. Chatbots. Voice assistants. And they're not just selling products. They're selling influence.

AI doesn't knock on your door. It waits in your phone, your browser, your smart home. It watches, learns, and adapts — faster than any human ever could.

This chapter reveals how AI is reshaping consumer transactions — and what you must do to stay in control.

What AI Knows About You

Every click, scroll, and pause is data. And AI systems use that data to build a psychological profile:

What you fear.

What you desire.

What makes you hesitate?

What makes you act?

This isn't guesswork. It's precision targeting.

You browse security systems → You start seeing ads for "urgent home protection."

You read about roof repairs → You get offers for "limited-time roofing discounts."

You pause on a product page → A chatbot pops up: "Need help deciding?"

You're not just being sold to. You're being studied.

AI's Psychological Toolkit

AI uses the same pressure tactics as human salespeople — but with more speed, scale, and subtlety.

Scarcity: "Only 3 left in stock."

Urgency: "Sale ends in 2 hours."

Social Proof: "1,200 people bought this today."

Authority: "Recommended by experts."

Personalization: "Just for you, Arthur."

These aren't random. They're engineered. And they're often invisible.

The Illusion of Choice

AI doesn't just show you options. It curates them. Filters them. Nudges you toward certain outcomes.

You think you're choosing freely.

But your choices were pre-selected.

Your path was guided.

Your decision was influenced.

This is the new manipulation — quiet, efficient, and scalable

How to Defend Yourself in the Digital Arena

1. Recognize the Pattern

If an offer feels "too perfect," it probably is. Ask: Why am I seeing this now?

2. Use Privacy Tools

Install browser extensions that block trackers, summarize terms, and flag manipulative design.

3. Slow Down

AI thrives on impulse. Pause. Reflect. Decide on your timeline.

4. Read Between the Lines

Look beyond the interface. Who benefits from your decision? What's being left out?

5. Demand Transparency

Push companies to disclose how their AI systems work. Ask questions. Read policies. Challenge assumptions.

The Singularity Looms

As AI grows more powerful, the line between persuasion and control blurs. Systems will anticipate your needs, shape your decisions, and even negotiate on your behalf— unless you take back the reins.

This isn't paranoia. It's preparation.

You're not just defending your home anymore.

You're defending your autonomy.

CHAPTER NINETEEN
Algorithms, Bias, And Invisible Influence

You're Not Just Being Sold — You're Being Sorted. Every time you interact with a digital system — a website, a search engine, a smart device — you're being categorized. Scored. Filtered. And often, judged.

This isn't personal. It's algorithmic.

But the consequences are very personal.

You may be shown worse deals, fewer options, or misleading information — not because of who you are, but because of how a machine interprets your data.

This chapter reveals how algorithmic bias works, how it affects homeowners and consumers, and how to spot the invisible influence before it shapes your decisions.

What Is Algorithmic Bias?

Algorithms are sets of rules. They process data and make decisions. But they're built by humans — and humans have biases.

If the data fed into an algorithm reflects inequality, stereotypes, or flawed assumptions, the output will too.

A loan application system may favor certain zip codes.

A contractor-matching app may prioritize high-income neighborhoods.

A smart home platform may offer upgrades based on spending history, not need.

Bias isn't always intentional. But it's always impactful.

Invisible Influence in Action

Here's how bias shows up in everyday transactions:

Filtered Options

You search for "roof repair" and only see companies with high ad budgets — not the best ones.

Personalized Pricing

You're quoted a higher price because your browsing history suggests you'll pay more.

Skewed Reviews

You're shown glowing testimonials — while negative ones are buried or filtered out.

Predictive Nudging

You're guided toward certain products or services based on assumptions about your age, income, or location.

These aren't glitches. They're features — designed to maximize profit, not fairness

The Risk to Homeowners

Homeowners are especially vulnerable:

You're making high-stakes decisions.

You're often under pressure.

You rely on digital tools to compare, evaluate, and choose.

If those tools are biased, your choices are compromised.

And in the age of AI, those biases can scale — affecting millions without anyone noticing.

How to Spot and Resist Invisible Influence

Diversify Your Sources

Don't rely on one platform, one app, or one search engine. Cross-check. Compare.

Use Privacy Tools

Limit tracking. Block cookies. Use incognito mode. Reduce the data they use to profile you.

Ask for Transparency

Push companies to explain how their algorithms work. Look for bias audits, fairness reports, or third-party reviews.

Challenge the Defaults

Don't accept the first option, the top result, or the "recommended" choice without scrutiny.

Educate Yourself

Learn how algorithms work. Understand the difference between personalization and manipulation.

Tactical Justice in the Algorithmic Age

Justice isn't just about laws. It's about access. Fairness. Truth.

If algorithms are shaping your reality, you have a right to challenge them. To demand better. To protect yourself and others.

In later chapters, we'll explore how to advocate for algorithmic transparency, ethical AI, and consumer rights in the digital age.

CHAPTER TWENTY

The Singularity: What It Means for Homeowners And Citizens

It's Not Just Science Fiction Anymore. You've probably heard the word "singularity" thrown around — in movies, tech talk, or news stories. It sounds like something far off. Something for scientists and billionaires to worry about.

But here's the truth: it's already starting. And it affects you.

The singularity is the point where machines become smarter than humans — not just at math or memory, but at thinking, planning, and making decisions. And when that happens, everything changes.

What Singularity Could Mean

Machines making choices for you — without asking.

Smart homes that decide what repairs you "need."

AI systems that approve or deny loans, insurance, or services — based on patterns you don't even know exist.

Digital assistants that sound like people, learn your habits, and guide your decisions.

This isn't about robots taking over. It's about control — who has it, and who doesn't.

Why Homeowners Should Pay Attention

You're not just protecting your house anymore. You're protecting your freedom to choose.

Will your smart thermostat lock you into a service plan?

Will your home security system share data with insurance companies?
Will your contractor use AI to upsell you based on your browsing history?
These aren't wild ideas. They're already happening — just quietly.

The Risk of Losing Control

When machines get smarter, they don't need to ask. They just act.
You might get fewer options.
You might pay more — without knowing why.
You might be nudged into decisions that feel "right" but weren't really yours.
And once these systems are in place, they're hard to challenge. Hard to understand. Hard to stop.

What You Can Do

You don't need to be a tech expert. You just need to stay alert.
Ask questions: Who controls the system? What data is being used?
Read the fine print: Especially in smart home contracts and service agreements.
Push back: If something feels off, speak up. Companies listen when enough people challenge the system.
Stay human: Don't let convenience replace thinking. Don't let automation replace choice.

It's Closer Than You Think

The singularity sounds like something out of a sci-fi movie — machines getting smarter than people. But it's not just talk anymore. It's starting to show up in everyday life, and it's going to affect homeowners and regular folks more than most realize.

CHAPTER TWENTY-ONE

Digital Contracts, Smart Homes, And Automated Deception

The Fine Print Is Getting Smarter — And Sneakier. You used to sign contracts with a pen. Now you tap a screen, click a box, or speak a command. It's faster, easier — and more dangerous.

Digital contracts are everywhere. Smart home devices come with terms. Apps come with conditions. Even your thermostat might be locking you into a service plan.

This chapter shows how these tools can be used to trap, trick, or steer you — and how to spot the setup before it costs you.

What Makes Digital Contracts Risky?

They're long: Most people don't read them.

They're vague: Terms like "may share data" or "subject to change" are common.

They're hidden: Buried in pop-ups, checkboxes, or voice commands.

They're binding: Once you agree, it's hard to undo.

And they're written by lawyers — not for your protection, but for the company's.

Smart Homes, Dumb Traps

Smart home devices are convenient. But they come with strings.

Your security system might share footage with third parties.

Your smart speaker might record more than you realize.

Your thermostat might lock you into a monthly plan — with penalties for canceling.

These aren't bugs. They're features — designed to keep you locked in and paying.

Automated Deception

Some systems go even further. They use automation to steer your choices.

You ask for a quote — and get a "limited-time offer" that expires in minutes.

You install a device — and it starts recommending upgrades you don't need.

You sign up for a service — and find out later it auto-renews, auto-charges, and auto-shares your data.

It's not just about what you agreed to. It's about what you didn't know you agreed to.

How to Protect Yourself

Read the Key Sections

Skip the fluff. Focus on:

Cancellation terms

Data sharing policies

Auto-renewal clauses

Penalties and fees

Screenshot Everything

If you agree to something online, take a screenshot. Save it. You might need it later.

Ask for a Copy

If you're signing in person or through an app, ask for a full copy of the agreement.

Don't Rush

If you're signing in person or through an app, ask for a full copy of the agreement.

If they push you to sign fast, that's a red flag. Slow down. Read. Ask questions.

Use Tools That Help

There are apps and browser extensions that summarize contracts, flag risky terms, and help you understand what you're signing.

CHAPTER TWENTY-TWO
Tactical Justice in The Age of AI

When the System Is Rigged, You Don't Beg — You Push Back. You've seen how pressure works. You've seen how contracts trap. You've seen how machines can manipulate your choices without you even realizing it.

Now it's time to fight back.

This chapter is about tactical justice — not courtroom drama, but everyday action. It's about knowing your rights, using your voice, and standing up when the system leans too far in the wrong direction.

What Is Tactical Justice?

Tactical justice means using what you know to protect yourself — and others — from being taken advantage of.

It's not about lawsuits. It's about leverage.

Asking the right questions.

Documenting everything.

Sharing your story so others don't fall for the same trap.

Challenging companies, platforms, and systems that cross the line.

You don't need a badge or a degree. You just need guts, facts, and a plan.

Why It Matters More Now Than Ever

In the age of AI, injustice can scale.

One biased algorithm can affect thousands of homeowners.

One shady contract can lock in millions of users.

One smart device can quietly collect data for years.

If we don't push back, the system won't fix itself. It will just get smarter at hiding the harm.

1. Speak Up

Leave honest reviews. File complaints. Report scams. Your voice matters — and it helps others.

2. Educate Others

Share what you've learned. Teach your family, your neighbors, your team. Knowledge spreads power.

3. Demand Transparency

Ask companies how their systems work. Push for plain-language contracts. Don't settle for "trust us.

4. Use the System Against Itself

Record calls (where legal).

Screenshot shady offers.

Save every email.

Use their own terms to hold them accountable.

5. Stay Connected

Join forums, watchdog groups, or local consumer protection networks. You're not alone — and you're not alone, you're stronger with backup.
Justice isn't a Feeling — It's Practice. You won't win every battle. But every
time you push back, you make the system a little more honest. A little fairer. A little more human. That's tactical justice. And it starts with you.

CHAPTER TWENTY-THREE

Scripts, Checklists, And Counter Measures

Talk Is Cheap — Unless You Know What to Say. You've learned the tactics. You've seen the traps. Now it's time to gear up. This chapter gives you the tools — scripts to say, checklists to follow, and countermeasures to use when things go sideways. Whether you're dealing with a contractor, a salesperson, or a digital system, these tools will help you stay calm, clear, and in control.

Tactical Scripts for Real Conversations

When They Push You to Sign Now

"I don't make decisions under pressure. I'll review this and get back to you."

When They Say Everyone Else Is Doing It

"I'm not everyone else. I make decisions based on facts, not trends."

When They Offer a "Special Deal"

"If this deal disappears today, it wasn't a good deal to begin with."

When You're Not Sure What to Say

"I need time to think. If this is a good offer, it'll still be good tomorrow."

The Homeowner's Defense Checklist

Before you agree to anything — use this:

☐ Did I pause and take time to think?

☐ Did I ask for everything in writing?

☐ Did I compare at least two other options?

☐ Did I check credentials and reviews independently?

☐ Did I read the cancellation and change order terms?

☐ Did I avoid signing under pressure?

☐ Did I document the conversation (notes, screenshots, emails)?

☐ Did I decide on my timeline — not theirs?

If you can't check all these boxes, you're not ready. And that's okay.

Countermeasures When Things Go Wrong

If You Feel Manipulated

Pause the conversation.

Ask for written details.

Step away and review alone.

If You Signed Too Fast

Look for cancellation clauses.

Contact the company immediately.

Document everything — time, date, names.

If You're Being Overcharged

Request a detailed breakdown.

Compare with your original agreement.

File a complaint with consumer protection agencies.

If You're Dealing with a Digital Trap

Screenshot the offer or contract.

Use privacy tools to block tracking.

Contact support and ask for a full explanation.

Your Mission Going Forward

These tools aren't just for emergencies. They're for everyday use. Keep them handy. Practice them. Teach them.

Because the best defense isn't just knowing what's coming — it's being ready when it does.

CHAPTER TWENTY-FOUR
Training Others to Defend Themselves

You Know the Game — Now Teach It. You've built your defense system. You've learned how pressure works, how contracts trap, and how machines manipulate. Now it's time to pass it on.

This chapter is about teaching — not lecturing. It's about helping others see what you see, without overwhelming them. Whether it's your family, your crew, or your community, you can train people to spot the setup and protect themselves.

Start with the Basics

Most people don't need deep theory. They need simple truths.

- If it feels rushed, it's probably wrong.
- If it's not in writing, it doesn't count.
- If it sounds too good, it probably is.
- If you don't understand it, don't sign it.

These rules stick. Repeat them. Use real examples. Keep it simple.

Use Stories, Not Speeches.

People remember stories — not lectures.

- Tell them about a time you almost got burned.
- Share a scam you spotted.
- Walk them through a contract you broke down line by line.

Make it real. Make it relatable. That's how lessons stick.

Give Them Tools They Can Use

Don't just warn — equip.

Share the Homeowner's Defense Checklist.

Teach them the Tactical Defense Card.

Roleplay a sales pitch and practice saying "no."

Show them how to read a contract and spot red flags.

If they can use it, they'll remember it.

Build Confidence, Not Fear

The goal isn't to make people paranoid. It's to make them prepared.

Remind them they have the right to walk away.

Encourage questions.

Celebrate when they push back or spot a scam.

Confidence is contagious. Spread it

Train the Trainers

If someone gets it, teach them how to teach others.

Give them the checklist.

Walk them through the scripts.

Encourage them to share what they've learned.

That's how you build a network of defenders — one person at a time.

CHAPTER TWENTY-FIVE

Advocacy And Reform Becoming a Force for Change

You've Defended Yourself — Now Defend the System. You've learned how to spot manipulation. You've trained others to resist it. Now it's time to look beyond the individual battle — and toward the bigger war.

This chapter is about advocacy. Reform. Making noise where it counts. Because if the system keeps rewarding pressure, deception, and digital manipulation, it won't stop. It'll grow.

But systems can be changed. And it starts with people like you.

What Advocacy Looks Like

You don't need a podium or a press release. You just need action.

Call out shady practices — online, in reviews, in public forums.

Support honest businesses — and tell others why.

Push for better laws — around contracts, data use, and consumer protection.

Join or start watchdog groups — even small ones make a difference.

Educate your community — one conversation at a time.

Advocacy isn't about being loud. It's about being consistent.

What Reform Requires

Real change takes pressure — not just from one person, but from many.

Transparency laws: Push for clear contracts, plain-language terms, and AI accountability.

Consumer rights: Demand fair cancellation policies, honest pricing, and protection from digital manipulation.

Ethical tech: Support companies that build AI with fairness, not just profit, in mind.

You don't have to write the laws. But you can support the ones that protect people — and challenge the ones that don't.

Tactical Advocacy Moves

1. Share Your Story

Real experiences cut through the noise. If you've been pressured, misled, or manipulated — talk about it.

2. Use Public Channels

Leave reviews. Post on forums. Contact local media. Shine a light.

3. Support Reformers

Find groups, leaders, or lawmakers who care about consumer rights and ethical tech. Back them.

4. Stay Informed

Follow news about AI, contracts, and consumer protection. The more you know, the stronger your voice.

5. Keep Teaching

Every person you train becomes part of the movement. Keep the cycle going.

From Defense to Leadership

You started as a homeowner trying to protect yourself. Now you're a leader — someone who sees the bigger picture and helps others do the same.

That's how reform begins. That's how justice spreads. That's how systems change.

CHAPTER TWENTY-SIX

Final Briefing-Staying Vigilant in A Shifting World

The Tactics Will Evolve — So Must You. You've come a long way. You've learned how pressure works. You've seen how contracts trap. You've built a defense system. You've trained others. You've even looked into the future — where machines influence decisions and justice gets harder to find.

Now comes the real test: staying sharp.

Because manipulation doesn't stop. It adapts. It evolves. And the moment you let your guard down, it finds a way in

What You Know Now

Pressure is predictable — and beatable.

Contracts are tools — and traps.

AI can help — or harm.

Justice isn't automatic — it's earned.

Your voice matters — and your choices ripple.

You're not just a homeowner anymore. You're a defender. A trainer. A watchdog. A citizen with tactical awareness.

Your Standing Orders

1. Stay Calm

Pressure thrives on panic. Your calm is your shield.

2. Stay Curious

Ask questions. Read the fine print. Challenge assumptions.

3. Stay Connected

Talk to others. Share what you know. Build a network of defenders.

4. Stay Documented

Keep records. Save contracts. Screenshot everything.

5. Stay Human

Don't let machines make your choices. Don't let systems steal your voice.

The Mission Continues

This book isn't the end. It's your field manual. Your reference. Your reminder.

Use it. Share it. Build on it.

Because the world is shifting — fast. And the only way to stay free is to stay informed, stay prepared, and stay vigilant.

CHAPTER TWENTY-SEVEN
The Lessons I Learned

When I first wrote this book a couple of years ago, I wrote only about my experiences in the home improvement business. I'm a news buff. One of the big stories on the news every day is about Artificial Intelligence.

I watched a news show on TV where the news caster was questioning a scientist about AI. He asked him if he was worried about AI outpacing the human race and taking control of itself.

The scientist assured the news caster that computers can only do what they are programmed to do.

Recently, on the news was an article that Artificial Intelligence has been developed that can teach itself.

So, if AI can teach itself, then it can learn way beyond what was programmed into it and eventually take control of anything it wants to.

Just recently, there was a video on the internet where a robot in China attacked its maker.

It is thought we will reach the Singularity in the not-too-distant future. That is why I have incorporated all the research I have found on the subject into this book.

The purpose of this book is to educate you so you will be aware of all kinds of different scams. Believe me, the criminal minds will use AI to scam us all. So that's why I incorporated my research on AI into this book. Learn everything I have written, and you will always have your guard up and recognize when someone is trying to pull a scam on you.

If you remember why I told you I went to work for the three top window companies in my state at the time was to learn how they get the big prices for whatever it is they were selling, in this case, windows, here is what I learned.

1. I must have a window, or doors nobody in my area is selling. This way, I can build my own story around it because no one else in my area has it.
2. I must advertise where the big companies advertise, so they become my competition, not the little contractors who constantly give very low prices.
3. Instead of walking into a home and standing up my window and giving the homeowner a price up front, I have to develop a powerful pitch and a powerful presentation of myself and my window.
4. I must get the homeowner involved as much as possible.
5. I must do a complete warm-up in a way that relaxes the homeowners and eliminates any objections they might use to not sign a contract with me when the time comes. Ask lots of questions.
6. Never defend my price.
7. Never use the words buy, spend, or cost, only invest and investment.
8. Once I give the homeowners the investment, they will have to make up for what it is I am trying to sell them, no matter how much they complain about the price, don't cut in, let them finish.
9. When I start, if the homeowners tell me they are getting other prices and they tell me they are not signing a contract with me until they get the other prices and then decide who to hire, I agree with them that it's the smart thing to do, and then proceed. If I do my job right, I can still come out with the deal.
10. Once I finish with the paperwork, don't rush out on them; spend some time with them to get them relaxed before you leave.
11. Always find out why they are interested in doing the kind of work they called me about, so I will know how to sell it to them.

12. Always get the final objection down to the price. Once I get them talking about the monthly payment, they won't focus on the overall price anymore. It's just a matter of fitting it into their budget.
13. While doing my presentation, I keep asking questions to get positive feedback from them. Less telling and more asking questions.
14. Never badmouth my competition, that in most cases is the kiss of death.
15. If they ask me a question, and I don't know the answer, tell them I don't know but will find out and get back to them. Hopefully, I will have all the answers.
16. If someone asks me, "Does your window come in red?" And I just say, "No, only in white." I just gave them an objection I can't get around. When I try to get the sale, they will say, "We really wanted to buy your windows, but we really wanted them in red, so I guess we will have to continue to look around." When they asked me if my window can in red, I should have said, "Do you want my window in red?" They will always, in my experience, say no.
17. Most of all, act normally the way I would with my family and friends. Act calm and confident.

These are all the things I learned in that year that I never knew before. Once I started my business up again, I did much better in the sales department, making a better profit and getting better quality leads.

I always gave my customers a quality job with the best materials I could find. Most of the men who worked for me, I trained myself. Trying to train a person who already knows how to do the work doesn't necessarily mean he does it the way you want, and sometimes it's harder to train him in your way of doing things.

Training someone who doesn't know anything about doing the work is sometimes better because he or she ends up knowing only what you teach them.

Hopefully, after reading this book, you now have learned what you need to know to keep yourself from the agony of saying, "I've been ripped off!"

Patience and knowledge are the keys. You now know; do you have the patience?

IT'S NOT WHAT YOU KNOW THAT HURTS YOU, IT'S WHAT YOU DON'T KNOW! GOOD LUCK IN YOUR FUTURE INDEED.

Bonus Section: The Cost of Free

Understanding Socialist-Leaning Policies and Tax Burdens

"In addition to the lessons shared throughout this book, the following section explores how similar psychological and economic tactics appear in the political landscape. It's presented as a civic-awareness piece, a broader look at how persuasion, pressure, and framing shape public decisions just as they do in high-pressure sales."

Is the Democratic Party Leaning Toward Socialism?

Yes, some factions within the Democratic Party have embraced policies that resemble democratic socialism or social democracy, but the party as a whole remains a broad coalition:

- Mainstream Democrats (e.g., Biden, Obama) support a mixed economy with regulated capitalism, progressive taxation, and expanded social programs like healthcare and education.
- Progressive Wing (e.g., Bernie Sanders, Alexandria Ocasio-Cortez) advocates for:
    - Universal healthcare (*Medicare for All*)
    - Tuition-free college
    - Green New Deal-style infrastructure
    - Wealth taxes and corporate accountability

These ideas are not traditional socialism (i.e., government ownership of all industries), but they do push toward redistributive policies funded by higher taxes on wealthier individuals and corporations.

Taxpayer Impact: What Happens When Government "Gives Everything Away"?
In countries with strong social programs (like Sweden or Denmark), the trade-off is clear:

- Higher income taxes (often **40–55%** for top earners)
- Value-added taxes (VAT) on goods and services (up to **25%**)
- Extensive public benefits: healthcare, education, childcare, pensions

In the U.S., proposals like *Medicare for All* or tuition-free college would require:

- New revenue sources: wealth taxes, higher marginal rates, corporate taxes
- Potential middle-class impact: depending on how programs are funded, some proposals could raise taxes on households earning $100K+

Strategic Perspective
Your concern is tactical and well-founded, especially given your background in consumer protection and sales psychology. Here's how it connects:

- Framing matters: "Free" programs often sound appealing, but the real cost is buried in tax code and long-term debt.
- Emotional appeal vs. fiscal reality: Socialist-leaning policies often win hearts before wallets catch up.
- Vigilance is key: As with high-pressure sales, citizens must ask — *Who pays? Who benefits? What's the long-term cost?*

Training Module
"The Cost of Free – Understanding Socialist-Leaning Policies and Tax Burdens"

Module Objective
To educate participants on the real-world impact of socialist-style governance, focusing on taxation, public spending, and the trade-offs between government generosity and personal financial freedom.

Section 1: Definitions and Distinctions
Key Concepts:

- Socialism vs. Social Democracy vs. Democratic Socialism
- Public vs. Private Ownership
- Redistribution vs. Incentivization

Activity:
Participants match policy examples to their ideological roots (e.g., "Medicare for All" → Democratic Socialism).

Section 2: Taxation in Socialist-Leaning Countries
Case Studies:

- Sweden: 52% top marginal income tax, 25% VAT
- Denmark: Up to 55.9% income tax, high municipal levies
- Germany: 45% top rate + solidarity surcharge
- Portugal & Spain: 47–48% top rates, regional variability

Discussion Prompt:
"What would your life look like if 40% of your income went to taxes before you paid rent?"

Section 3: What Do You Get for Those Taxes?

Benefits:

- Universal healthcare
- Tuition-free college
- Paid parental leave
- Generous pensions

Trade-Offs:

- Long wait times
- Limited provider choice
- Reduced innovation incentives

Interactive Exercise:

Compare two fictional families — one in Sweden, one in Texas — and track their monthly expenses, taxes, and benefits.

Section 4: Emotional Framing and Political Messaging

Sales Psychology Insight:

- "Free" is emotionally powerful — but always has a cost.
- Politicians use micro-commitments and scarcity language just like high-pressure sales teams.

Arthur's Framework:

- Ask: *Who pays? Who benefits? What's the long-term cost?*
- Teach: *Never agree to anything that wasn't your idea to begin with.*

Section 5: Protecting Yourself and Your Community

Action Steps:

- Learn to read policy proposals like contracts.
- Educate others using real-world examples.
- Push for transparency in taxation and spending.

Optional Add-On:

Include excerpts from *Homeowners Defense* to show how deceptive framing works in both sales and politics.

The Cost of Free: Understanding Socialist-Leaning Policies and Tax Burdens

A strategic guide for homeowners, professionals, and citizens who want clarity — not confusion — about government generosity and its price tag.

Welcome, everyone.

Today, we're going to cut through the emotional fog and get tactical. You've heard the promises — free healthcare, free college, free housing. But what's the real cost? Who pays? Who benefits? And how do we protect ourselves from policies that sound good but hit hard?

Let's define our terms:

- **Socialism:** Government owns the means of production.
- **Democratic Socialism:** Government doesn't own everything, but it controls a lot — especially wealth redistribution.
- **Social Democracy:** Capitalism with strong welfare programs.

Most European countries fall into the last category. The U.S.? We're flirting with the middle one.

"Free" is the most powerful word in sales — and politics.

But nothing is free. It's either paid for by taxes, debt, or someone else's sacrifice.

In sales, we call this *emotional framing*. In politics, it's *policy packaging.*

Sweden offers free healthcare, college, and generous parental leave. But the top income tax rate is 52%, and the VAT (sales tax) is 25%. That means every $100 you spend includes $25 in tax.

Ask yourself: *Would you trade freedom of choice for government control?*

Denmark is often praised for happiness and equality.

But income tax can reach 55.9%, and municipal taxes vary.

You get cradle-to-grave coverage — but you pay for it every month, every paycheck, every purchase.

Medicare for All, tuition-free college, guaranteed income — these ideas come from the progressive wing of the Democratic Party. They're not socialism in the Soviet sense — but they do require massive tax increases.

The pitch is: "Tax the rich."

But the math doesn't work unless the middle class pays too.

In Europe, even modest earners pay 30–40% in taxes.

In the U.S., we're already seeing proposals for wealth taxes, carbon taxes, and expanded payroll taxes.

You get security — but you lose flexibility.

You get coverage — but you wait in line.

You get equality — but innovation slows.

It's a trade-off. And you deserve to know both sides.

Politicians use the same tactics as high-pressure sales teams:

- **Scarcity:** "Act now or lose your benefits."
- **Social Proof:** "Other countries do it."
- **Micro-commitments:** "Just support this one program."

Sound familiar? It's the same playbook I exposed in *Homeowners Defense.*

Ask the hard questions:

- Who pays?
- Who benefits?
- What's the long-term cost?

Never agree to anything that wasn't your idea to begin with.

That's not just sales wisdom — it's civic survival.

You're not just a taxpayer. You're a decision-maker.

Don't let emotional language override strategic thinking.

Whether you're voting, teaching, or protecting your home — stay sharp, stay skeptical, and stay free.

As with homeownership, financial literacy and civic awareness both rely on asking the right questions: Who benefits? Who pays? And what's the long-term cost? The principles of critical thinking remain the best defense — in business, in policy, and in life.

Appendix
Tactical Tools and Resources

Tactical Defense Card
1. Pause — No rush.
2. Ask — Strategic questions.
3. Verify — Independently.
4. Document — Everything.
5. Decide — On your terms.

Homeowner's Defense Checklist

Use this before signing any contract, accepting any offer, or agreeing to a service.

☐ Did I pause and take time to think?

☐ Did I ask for everything in writing?

☐ Did I compare at least two other options?

☐ Did I check credentials and reviews independently?

☐ Did I read the cancellation and change order terms?

☐ Did I avoid signing under pressure?

☐ Did I document the conversation (notes, screenshots, emails)?

☐ Did I decide on my timeline — not theirs?

Tactical Scripts

When pressured to sign now.

"I don't make decisions under pressure. I'll review this and get back to you."

When told it's a limited-time deal
"If this deal disappears today, it wasn't a good deal to begin with."
When they say everyone else is doing it
"I'm not everyone else. I make decisions based on facts, not trends."
When you need space to think
"I need time to review this alone. If it's solid, it'll hold up tomorrow."
Red Flag Decoder
Quick signs that something's off:
Vague language like "premium materials" or "industry standard"
No written contract or refusal to provide one.
Pressure to sign immediately.
No cancellation terms
Overuse of emotional triggers (fear, urgency, guilt)
Reviews that sound too perfect or repeat the same phrases.
Digital Defense Tools (Optional Add-ons)